# 石油科技英语系列教程

丛书主编 ◎ 江淑娟 吴松林

# Unconventional Oil and Gas Development

# 非常规油气开发

王利国 孙彩光 刘典忠 ◎ 编

English

石油工业出版社

## 内 容 提 要

　　本书是非常规油气开发相关知识的英语学习教程,主要包括非常规天然气介绍、油气产业、非常规天然气开发、非常规油气开发技术、生产对环境的影响、非常规油气市场、非常规油气开发宝典等内容。

　　本书适合石油相关专业的师生和石油系统员工学习石油科技英语之用。

**图书在版编目(CIP)数据**

非常规油气开发/王利国,孙彩光,刘典忠编.
北京:石油工业出版社,2016.9
石油科技英语系列教程
ISBN 978 – 7 – 5183 – 1427 – 0

Ⅰ. 非…
Ⅱ. ①王…②孙…③刘…
Ⅲ. 油气田开发 – 英语 – 教材
Ⅳ. H31

中国版本图书馆 CIP 数据核字(2016)第 190054 号

出版发行:石油工业出版社
　　　　(北京安定门外安华里 2 区 1 号　　100011)
　　　　网　　址:www. petropub. com
　　　　编辑部:(010)64253574　　图书营销中心:(010)64523633
经　　销:全国新华书店
印　　刷:北京中石油彩色印刷有限责任公司

2016 年 9 月第 1 版　2016 年 9 月第 1 次印刷
710×1000 毫米　开本:1/16　印张:20.75
字数:365 千字

定价:58.00 元
(如出现印装质量问题,我社图书营销中心负责调换)

# 《石油科技英语系列教程》
# 编 委 会

**丛书主编:** 江淑娟　吴松林

**编写人员:** (按姓氏笔画排序)

| | | | | |
|---|---|---|---|---|
| 马焕喜 | 王　红 | 王　琪 | 王　超 | 王化雪 |
| 王利国 | 历冬风 | 白雪晴 | 乔洪亮 | 朱年红 |
| 刘　葆 | 刘立安 | 刘亚丽 | 刘江红 | 刘杰秀 |
| 刘典忠 | 刘佰明 | 江淑娟 | 安丽娜 | 闫柯菲 |
| 孙彩光 | 苏亚杰 | 李　冰 | 李　健 | 李　超 |
| 李　滨 | 吴松林 | 谷秋菊 | 张　曦 | 张雪梅 |
| 陈　默 | 范金宏 | 卓文娟 | 赵　巍 | 姚坤明 |
| 秦艳霞 | 郭玉海 | 唐俊莉 | 徐国伟 | 高苏彤 |
| 康国强 | 隋　岩 | 韩福勇 | 葛多虹 | 董金玉 |
| 惠良虹 | | | | |

# 前　言

　　全球石油资源分布、生产及消费三者间存在着严重的地区失衡,中东和亚太是失衡最严重的地区,中东地区严重供过于求,亚太地区严重供不应求。因此能源行业出现了全球化发展趋势,能源国际间的交流与合作日益密切。为保证中国能源安全,中国石油和石化行业国际化和本土化发展势在必行。中国油气企业正在积极进行海外业务拓展,了解资源地区的文化背景、经济发展状况、能源开发政策以及掌握其石油地质结构、油气成藏条件、开发和炼制技术等将有利于我们对资源地区的油气开发和炼制,更有力地支持中国经济的快速发展。

　　自 1993 年起,为了满足石油院校和石油职工专业英语教材的严重匮乏,本丛书主编陆续在黑龙江人民出版社、黑龙江科技出版社、石油工业出版社等出版了系列专业石油英语教科书,积累了一定的编写经验、培训经验和图书项目导向经验。20 年过去了,石油行业也发生了巨大的变化,新油气资源不断发现,开采与炼制等技术不断更新,海外合作区域也不断拓宽。为了适应新形势,我们经过不懈努力通过了中国石油天然气集团公司图书出版立项,开始编写一套更大规模的《石油科技英语》丛书,既包括以石油上、中、下游生产技术为主线的硬科学部分,也包括世界主要石油资源国的经济、贸易和文化等软科学部分,目的是使读者奠定通向世界石油领域话语权的语言基础。

　　我们深感责任重大,从中国石油大学、东北大学、东北石油大学、西安石油大学及各油田石油地质研究院、设计院等单位聘请有关专家学者,确定编写提纲,搜集资料。在选材上,注重内容的系统性,争取覆盖本领域主要内容;语言方面,注意遴选突出科技英语语言特点的语段和篇章,并对语言使用方法做详尽解释,以英语基础知识和基本技能的培养为主。为降低学习难度,为每篇课文还配写了汉语译文,以提高学生的石油科技英语阅读、翻译及写作能力。

　　《非常规油气开发》共分七章,从非常规天然气介绍、油气产业、非常规天然气开发、非常规油气开发技术、生产对环境的影响、非常规油气市场、非常规油气开发宝典等方面介绍了非常规油气开发相关知识。在每一章内，为了能够让学习者更好地掌握课义内容，每一章均列出导读、相关的专业词汇及词组翻译、重点句子讲解，并跟进与重点内容相关的问题，读者在回答问题的同时，可以巩固对课文的理解，进而掌握相关专业知识。

《非常规油气开发》分册课文及译文由中国石油大学(北京)的王利国编写，第一章至第三章课文讲解及练习由中国石油大学(华东)的孙彩光编写，第四章至第七章课文讲解及练习由中国石油大学(华东)的刘典忠编写。蒋梦阳和朱文静帮助做了大量的资料查找和整理工作，在此一并表示感谢。

由于作者水平有限，且书中内容涉及面较广，可供参考、供鉴的资料不多，书中难免出现不足之处，敬请专家和读者批评指正。

丛书主编　江淑娟　吴松林
2015 年 4 月

# Contents

**Chapter 1  Introduction to Unconventional Oil and Gas** ·············· (1)

  1.1  Unconventional Gas Resources ···················· (1)

  1.2  Shale Gas ········· (9)

  1.3  Tar Sands ········· (17)

  1.4  Natural Gas Liquids ·············· (24)

**Chapter 2  Oil and Gas Industry** ··················· (33)

  2.1  Oil and Gas Industry Structure ··············· (33)

  2.2  Shale Gas Employment Outlook ············· (41)

  2.3  Exploration and Production Industry Trends ··············· (49)

  2.4  Oilfield Services Industry Trends ················· (57)

**Chapter 3  Unconventional Gas Development** ··············· (67)

  3.1  Shale and Tight Gas Development (1) ···················· (67)

  3.2  Shale and Tight Gas Development (2) ···················· (75)

  3.3  Coalbed Methane Development ·················· (83)

  3.4  Coalbed Methane Production and Effects on Groundwater ········· (91)

**Chapter 4  Unconventional Oil and Gas Production Technologies**

  ·················· (99)

  4.1  Exploration ················· (99)

  4.2  Well Drilling ··············· (108)

  4.3  Horizontal Drilling and Well Logging Methods ················· (116)

  4.4  Well Completion ··············· (125)

**Chapter 5  Production Impact on the Environment** ·············· (134)

  5.1  Water Use in Fracturing ··············· (134)

  5.2  Fracturing Fluid Additives ················· (142)

  5.3  Waste Water Treatment and Disposal ··············· (151)

5. 4　Greenhouse-gas Emissions ················································· (159)

**Chapter 6　Unconventional Oil and Gas Market** ······················· (168)

6. 1　Production Costs and Price of Natural Gas ······················· (168)

6. 2　Mineral Licensing and Leasing ······································ (176)

6. 3　Market Demand Drivers ············································· (185)

6. 4　Market Demand Inhibitors ·········································· (194)

**Chapter 7　Golden Rules for Unconventional Oil and Gas Development**

················································································· (203)

7. 1　Drilling Site Construction and Regulation ······················· (203)

7. 2　Environmental Impact Reduction ································· (210)

7. 3　Social Responsibilities ············································· (218)

**Chinese Translation and Key to Exercises** ······························ (227)

**References** ······································································ (324)

# Chapter 1　Introduction to Unconventional Oil and Gas

## 1.1　Unconventional Gas Resources

### ⌗ Guidance to Reading

*One of universally accepted definitions for unconventional oil and gas is that they are extracted using techniques other than the conventional method. In terms of the chemical composition, unconventional oil and gas are identical to conventional. Conventional oil includes crude oil, natural gas liquids and condensate liquids. Unconventional oil consists of more liquid sources including oil sands, extra heavy oil and shale oil, which need advanced technology to be extracted. Unconventional gas is found in highly compact rock or coalbeds and requires a specific set of production techniques. In general unconventional oil and gas are not easier and cheaper to produce. As economic and technological conditions evolve, resources hitherto considered unconventional can migrate into the conventional category. The categories of "conventional" and "unconventional oil and gas" will not remain fixed.*

### ⌗ Text

**Unconventional** gas refers to a part of the gas resource base that has traditionally been considered difficult or costly to produce. Here we focus on the three main **categories** of unconventional gas: shale gas, tight gas and coalbed methane.

**Shale gas** is natural gas contained within a commonly occurring rock classified as shale. Shale formations are characterized by low **permeability**, with more limited ability of gas to flow through the rock than is the case with a **conventional reservoir**. These formations are often rich in organic matter and, unlike most **hydrocarbon** reservoirs, are typically the original source of the gas, i. e. shale gas is gas that has remained **trapped** in, or close to, its source rock.

**Coalbed methane**, also known as coal **seam** gas, is natural gas contained in coalbeds. Although extraction of coalbed methane was initially undertaken to make mines safer, it is now typically produced from non-mineable coal seams.

**Tight gas** is a general term for natural gas found in low permeability formations. Generally, we classify as tight gas those low permeability gas reservoirs that cannot produce economically without the use of technologies to stimulate flow of the gas towards the well, such as **hydraulic fracturing**. Tight gas is often a poorly defined category with no clear boundary between tight and conventional, nor between tight gas and shale gas.

Although the development cycle for unconventional gas and the technologies used in its production have much in common with those used in other parts of the upstream industry, unconventional gas developments do have some distinctive features and requirements, particularly in relation to the perceived higher risk of environmental damage and adverse social impacts. This helps to explain why the issue of unconventional gas exploitation has generated so much controversy.

The main reason for the potentially larger environmental impact of unconventional gas operations is the nature of the resources themselves: unconventional resources are less concentrated than conventional deposits and do not give themselves up easily. They are difficult to extract because they are trapped in very tight or low permeability rock that **impedes** their flow. Since the resources are more diffuse and difficult to produce, the scale of the industrial operation required for a given volume of unconventional output is much larger than for conventional production. This means that drilling and production activities can be considerably more invasive, involving a generally larger environmental **footprint**.

One feature of the greater scale of operations required to extract unconventional gas is the need for more wells. Whereas **onshore** conventional fields might require less than one well per ten square kilometers, unconventional fields might need more than one well per square kilometer ( $km^2$ ), significantly intensifying the impact of drilling and completion activities on the environment and local residents.

Another important factor is the need for more complex and intensive preparation for production. While hydraulic fracturing is already used on occasions to stimulate conventional reservoirs, tight gas and shale gas developments almost always require the use of this technique in order to generate adequate flow rates into the well. The same technique is also often used, **albeit** less frequently, to produce coalbed methane. The associated use and release of water gives rise to a number of environmental concerns, including depletion of freshwater resources and possible contamination of surface water and aquifers.

The production of unconventional gas also contributes to the atmospheric concentration of greenhouse gases and affects local air quality. In some circumstances, unconventional gas production can result in higher **airborne** emissions of methane, a potent greenhouse gas, of **volatile organic compounds** (VOCs) that contribute to smog formation, and of carbon dioxide ($CO_2$) (from greater use of energy in the production process, compared with conventional production). Just how much greater these risks may be is uncertain: it depends critically on the way operations are carried out. On the other hand, there are potential net benefits from unconventional gas production, to the extent that, having been produced and transported to exacting environmental standards, it leads to greater use of gas instead of more carbon-intensive coal and oil.

In addition to the smaller recoverable hydrocarbon content per unit of land, unconventional developments tend to extend across much larger geographic areas. Moreover, areas with high unconventional potential are not always those with a strong or recent tradition of oil and gas industry activity; they are not necessarily rich in conventional **hydrocarbons** and in some cases there may have been little or no recent hydrocarbon production. This tends to **exacerbate** the problem of public acceptance.

### *Unconventional gas production and earthquake risks*

There have been instances of earthquakes associated with unconventional gas production, for example the case of the Cuadrilla shale gas operations near Blackpool in the United Kingdom, or a case near Youngstown, Ohio, in the United States, which has been provisionally linked to injection of waste water, an operation that is similar in some respects to hydraulic fracturing. The registered earthquakes were small, of a **magnitude** of around two on the Richter scale, meaning they were **discernible** by humans but did not create any surface damage.

Because it creates cracks in rocks deep beneath the surface, hydraulic fracturing always generates small **seismic** events; these are actually used by petroleum engineers to monitor the process. In general, such events are several orders of magnitude too small to be detected at the surface: special observation wells and very sensitive instruments need to be used to monitor the process. Larger seismic events can be generated when the well or the fractures happen to intersect, and reactivate an existing **fault**. This appears to be what happened in the Cuadrilla case.

Hydraulic fracturing is not the only anthropogenic process that can trigger small earthquakes. Any activity that creates underground stresses carries such a risk. Examples linked to construction of large buildings, or dams, have been reported.

**Geothermal wells** in which cold water is circulated underground have been known to create enough thermally-induced stresses to generate earthquakes that can be sensed by humans. The same applies to deep mining. What is essential for unconventional gas development is to survey carefully the geology of the area to assess whether deep faults or other geological features present an enhanced risk and to avoid such areas for fracturing. In any case, monitoring is necessary so that operations can be suspended if there are signs of increased seismic activity.

## Words and Expressions

| | |
|---|---|
| unconventional | 非常规的,非传统的 |
| category | 类别,分类 |
| permeability | 渗透率 |
| hydrocarbon | 碳氢化合物 |
| trap | 使……受限制,设陷阱;圈闭 |
| seam | 缝,接缝(夹、薄、矿、煤)层 |
| impede | 阻止,防碍 |
| foofpnnt | 足迹,脚印 |
| albeit | 虽然,即便 |
| airborne | 空气传播的,风媒的 |
| exacerbate | 使加剧,使恶化 |
| magnitude | 震级,量级 |
| discernible | 可辨别的,可识别的 |
| seismic | 地震的,因地震引起的 |
| fault | 断层 |

## Phrases and Expressions

| | |
|---|---|
| shale gas | 页岩气 |
| conventional reservoir | 常规油气藏 |

| coalbed methane | 煤层气 |
| tight gas | 致密气 |
| hydraulic fracturing | 水力压裂 |
| volatile organic compounds（VOCs） | 挥发性有机物 |
| geothermal well | 地热井 |

## Language Focus

1. Shale formations are characterized by low permeability, with more limited ability of gas to flow through the rock than is the case with a conventional reservoir.

（参考译文：页岩层的特点是渗透率低，与常规油藏相比，气体在页岩中的流动能力较差。）

本句中 be characterized by 为固定搭配，意思是"特点是……"。with 引导复合结构做状语，修饰整个句子，其中 than 引导比较状语从句，其后省略了 what，表示和常规气藏中天然气的流动能力相比。

2. Generally, we classify as tight gas those low permeability gas reservoirs that cannot produce economically without the use of technologies to stimulate flow of the gas toward the well, such as hydraulic fracturing.

（参考译文：我们把低渗透率气藏归类为致密气，这些低渗透率气藏如果不使用水力压裂等增产措施使气体流向井底，就不能使生产经济可行。）

本句主句可看做是：we classify those low permeability gas reservoirs as tight gas，因为 classify 的宾语 those low permeability gas reservoirs 后面还有一个 that 引导的定语从句，连在一起使宾语过长，因此把宾语补足语 as tight gas 提到了前面。

3. Since the resources are more diffuse and difficult to produce, the scale of the industrial operation required for a given volume of unconventional output is much larger than for conventional production.

（参考译文：由于这些资源比较分散、难以开采，所以要生产一定量的非常规天然气，其生产规模要远大于常规天然气开发。）

本句 since 引导原因状语从句；主句主谓结构是：the scale of... is much larger than...，其中含有一个 than 引导的比较状语从句，从句中省略了 the scale of the industrial operation required for the given volume of output，主句主语后面有过去分词短语 required for 作其后置定语。

4. On the other hand, there are potential net benefits from unconventional gas production, to the extent that, having been produced and transported to exacting environmental standards, it leads to greater use of gas instead of more carbon-intensive coal and oil.

（参考译文：另一方面，非常规天然气的开采会带来潜在的净收益，从这个意义上讲，严格按照环境标准进行开采和运输的天然气将被广为使用，取代含碳量更高的煤和石油。）

本句为复合句，主句是：there are potential net benefits from unconventional gas production。从句由 that 引导，从句主谓结构是：it leads to greater use of...。having been 是现在分词短语做状语，表示先一步发生的动作。

5. Geothermal wells in which cold water is circulated underground have been known to create enough thermally-induced stresses to generate earthquakes that can be sensed by humans.

（参考译文：因为冷水循环到地下进入地热井中，所以地热井产生热诱导应力，足以引发人类可以感觉到的地震。）

in which 引导定语从句修饰 Geothermal wells，指代 in the geothermal wells，that 引导的定语从句修饰 earthquakes。

## Reinforced Learning

### Ⅰ. Answer the following questions for a comprehension of the text.

1. What is unconventional gas?

2. What is the main reason for the potentially larger environment impact of unconventional gas operations?

3. How do unconventional gas operations affect the environment?

4. Why do we say that drilling and production activities can be much more invasive in unconventional gas operations?

5. How can we control the earthquake risk generated from unconventional gas production?

### Ⅱ. Multiple choice: choose the correct one from the alternative answers to give the exact meaning of the words.

1. Shale formations are characterized by low permeability, with more limited ability of gas to flow through the rock than is the case with a conventional reservoir.

A. perviousness   B. breathability   C. transferability   D. conductivity

2. Although extraction of coal-bed methane was initially undertaken to make mines safer, it is now typically produced from non-mineable coal seams.

A. use          B. application     C. discovery       D. exploitation

3. Unconventional gas developments do have some distinctive features and requirements.

A. important     B. unusual       C. apparent        D. significative

4. They are difficult to extract because they are trapped in very tight or low permeability rock that impedes their flow.

A. slows         B. interrupts     C. stops           D. accelerates

5. This means that drilling and production activities can be considerably more invasive, involving a generally larger environmental footprint.

A. influential    B. aggressive     C. violent         D. ferocious

6. Whereas onshore conventional fields might require less than one well per ten square kilometers, unconventional fields might need more than one well per square kilometer ($km^2$), significantly intensifying the impact of drilling and completion activities on the environment and local residents.

A. enhancing     B. exacerbating   C. impairing       D. enlarging

7. The same technique is also often used, albeit less frequently, to produce coalbed methane.

A. but          B. then          C. although        D. when

8. A case near Youngstown, Ohio, in the United States, which has been provisionally linked to injection of waste water.

A. firstly       B. briefly        C. finally          D. ultimately

9. Hydraulic fracturing is not the only anthropogenic process that can trigger small earthquakes.

A. man-made     B. contrived      C. industrialized   D. spontaneous

10. In any case, monitoring is necessary so that operations can be suspended if there are signs of increased seismic activity.

A. stopped       B. abolished      C. canceled        D. admonished

**III. Multiple choice: read the four suggested translations and choose the best answer.**

1. Generally, we classify as tight gas those low permeability gas reservoirs that cannot produce economically without the use of technologies to stimulate

flow of the gas towards the well, such as hydraulic fracturing.

    A. 刺激        B. 鼓励        C. 加快        D. 激发

2. This helps to explain why the issue of unconventional gas exploitation has generated so much controversy.

    A. 争议        B. 反对        C. 对立        D. 辩论

3. Since the resources are more diffuse and difficult to produce, the scale of the industrial operation required for a given volume of unconventional output is much larger than for conventional production.

    A. 不同的        B. 分散的        C. 特别的        D. 难利用的

4. The associated use and release of water gives rise to a number of environmental concerns, including depletion of freshwater resources and possible contamination of surface water and aquifers.

    A. 损耗        B. 下降        C. 开采        D. 利用

5. This tends to exacerbate the problem of public acceptance.

    A. 激化        B. 增加        C. 影响        D. 规避

### Ⅳ. Put the following sentences into Chinese.

1. The low permeability gas reservoirs cannot produce economically without the use of technologies to stimulate flow of the gas towards the well, such as hydraulic fracturing.

2. They are difficult to extract because they are trapped in very tight or low permeability rock that impedes their flow.

3. One feature of the greater scale of operations required to extract unconventional gas is the need for more wells.

4. The registered earthquakes were small, of a magnitude of around two on the Richter scale, meaning they were discernible by humans but did not create any surface damage.

5. In any case, monitoring is necessary so that operations can be suspended if there are signs of increased seismic activity.

### Ⅴ. Put the following paragraphs into Chinese.

1. In addition to the smaller recoverable hydrocarbon content per unit of land, unconventional developments tend to extend across much larger geographic areas. Moreover, areas with high unconventional potential are not always those with a strong or recent tradition of oil and gas industry activity; they are not

necessarily rich in conventional hydrocarbons and in some cases there may have been little or no recent hydrocarbon production. This tends to exacerbate the problem of public acceptance.

2. Because it creates cracks in rocks deep beneath the surface, hydraulic fracturing always generates small seismic events; these are actually used by petroleum engineers to monitor the process. In general, such events are several orders of magnitude too small to be detected at the surface: special observation wells and very sensitive instruments need to be used to monitor the process. Larger seismic events can be generated when the well or the fractures happen to intersect, and reactivate an existing fault.

# 1.2    Shale Gas

## Guidance to Reading

*Shale gas offers huge potential, accounting for the majority of the world's recoverable unconventional gas. Because shales ordinarily have insufficient permeability to allow significant fluid flow to a well bore, most shales are not commercial sources of natural gas. However, shale gas in the United States is rapidly increasing as an available source of natural gas. Led by new applications of hydraulic fracturing technology and horizontal drilling, development of new sources of shale gas has offset declines in production from conventional gas reservoirs, and has led to major increases in reserves of US natural gas.*

## Text

The total global shale gas resource base is enormous. Estimated to **exceed** 18,775 trillion cubic feet (**Tcf**), the resource potential of gas from shales exceeds that of all total world conventional resources combined. To further lend perspective to the magnitude of estimated global shale gas resources, current global natural gas consumption is estimated at about 110 Tcf annually. Even if only 25 percent of this total resource base becomes technically recoverable, shale gas alone can meet global natural gas demand for over 40 years at current consumption levels.

Gas shales are organic-rich formations with total organic carbon **accounting for** up to 25 percent of their composition by weight. Existing three to six kilometers underground, they are typically found at much greater subsurface depths than conventional gas **reservoirs**. Shales have low permeability matrices and act as both the **source** and **reservoir rock** for gas. Further, they are **horizontally** oriented underground. These characteristics of shale resources add significant complexity and cost to gas exploration and production and **distinguish** shale **from** conventional natural gas resources.

### Geology of Shale Gas Resources

Shale is a **sedimentary** rock predominantly **comprised of consolidated** clay-sized particles. Shales are deposited as mud in low-turbulence depositional environments, such as **tidal flats** and deep water basins, where the fine-grained clay particles fall out of **suspension**. Organic matter such as algae − , plant − , and animal-derived organic debris is often **trapped** during sediment deposition.

The naturally **tabular** clay grains tend to lie flat as the sediments accumulate. With years of deposition, these sediments become highly compacted and **solidify** into thinly layered shalerock. The combination of compaction and **lamination** result in a **formation** with extremely limited permeability. Over millions of years, the organic material trapped in shale becomes compressed and heated, forming hydrocarbons including natural gas. Figure 1. 1 shows a typical shale outcrop which reveals the natural bedding planes, or layers, of shale and **nearvertical** natural fractures that can cut across the naturally horizontal bedding planes.

The natural gas is trapped in the pore spaces between individual mineral grains, in the natural fractures, and is also chemically adsorbed onto organic matter within the shale. The low **matrix** permeability of shale means that gas trapped cannot move easily within the rock and typical shales can have matrix permeability on the order of 0. 01 to 0. 00001 millidarcies.

**Figure 1. 1   Shale Outcrop**

Thus, shale gas is both produced and stored within the shale bed. Such formations are therefore viewed as both source rocks and reservoirs for natural gas and are often called gas shales.

### Chemical Composition of Shale Gas

Natural gas is a combination of hydrocarbon gases **consisting** primarily **of** methane ( $CH_4$ ) and, to a lesser extent, butane, ethane, propane, and other gases. It is odorless, colorless, and, when **ignited**, releases a significant amount of energy. Natural gas is considered "dry" when it is almost pure methane, having had most of the other commonly associated hydrocarbons removed. When other hydrocarbons are present, the natural gas is "wet". Although some shale formations do produce wet gas, shale gas **is** typically a dry gas primarily **composed of** 90 percent or more methane.

Gas processing removes one or more components from produced gas to prepare it for use in residential or industrial applications. Common components removed to meet pipeline, safety, environmental, and quality specifications include hydrogen **sulfide**, carbon dioxide, nitrogen, heavy hydrocarbons, and water. The technique employed to process the gas varies with the components to be removed as well as with the properties of the gas stream ( e. g. , temperature, pressure, composition, flow rate ). Since the composition of shale gas is not the same, the gas processing requirements for shale gas can vary from play to play, and even among wells within the same play.

### *Common Characteristics of Economically Viable Shale Gas Resources*

Shales that host economic quantities of gas have a number of common properties. In general, these formations are:

- Rich in organic material, with total organic carbon comprising up to 25 percent by weight.

- Mature petroleum source rocks found at subsurface depths of three to six kilometers with temperature regimes of 100 to 200 degrees Celsius ( also called the **thermogenic** gas window ); the high heat and pressure at these depths have converted organic material to natural gas.

- Sufficiently brittle such that they can be fractured and rigid enough to maintain open fractures.

The specific characteristics of gas shales have been found to vary both among and within economically **viable** basins. For instance, their thermal maturity and chemical composition can vary such that hydrocarbon content ranges from desirable methane to higher molecular weight hydrocarbon gases and liquids. Additional variations that can exist within and between shale formations include water content, reservoir pressure, thickness, permeability, and porosity. The absence or presence of geologic features such as faults is another factor which distinguishes shales from one another. Geologists and engineers must consider all of these factors when assessing the potential productivity of a particular reservoir and developing well plans.

Considerable advances in exploration, **drilling**, and production technologies have made accessing and recovering gas from low-permeability shales more economically viable over the past two decades. Advances in **horizontal drilling** have improved drilling efficiency and gas production rates. These technologies enable wells to reach and accurately follow long, narrow, horizontal **pay streaks**. This not only promotes increased reservoir contact, but also provides access to gas reservoirs that were previously difficult or impossible to reach, such as those that run under heavily populated areas or geologic features like lakes.

Promoting gas flow from low-permeability shales has been further facilitated by advances in reservoir **stimulation** techniques, specifically **hydraulic fracturing**. The process is essentially a controlled fracturing of shale rock to create conductive paths for gas to flow to the wellbore. **Fracturing fluids** have evolved to address the varied characteristics and resultant challenges of different shales from low-pressure, clay intensive and water-sensitive shales to highpressure formation with greater concentrations of brittle rock. Multi-stage fracturing tools have increased the speed and effectiveness with which long horizontal **laterals** can be fractured.

## Words and Expressions

| | |
|---|---|
| exceed | 超越,超过 |
| Tcf | 一万亿立方英尺 |
| reservoir | 油藏 |
| horizontally | 水平地,横地 |
| sedimentary | 沉淀的 |
| consolidate | 巩固,加强 |
| suspension | 悬浮,暂停 |
| trap | 圈闭 |
| tabular | 板状的;薄层的;表格式的;平的 |
| solidify | 凝固,团结 |
| lamination | 制成薄板,叠片结构 |
| formation | 形成;地层 |

| | |
|---|---|
| nearvertical | 接近垂直的 |
| matrix | 基质,脉石 |
| ignite | 燃烧,点燃 |
| sulfide | 硫化物 |
| thermogenic | 产热的,生热的 |
| viable | 可行的,生存的 |
| stimulation | 刺激;激发;促进 |
| lateral | 侧部,横向的 |

## Phrases and Expressions

| | |
|---|---|
| account for | 占 |
| source rock | 烃源岩 |
| reservoir rock | 储集岩 |
| distinguish from | 区别;辨别 |
| tidal flat | 潮滩 |
| be comprised of | 由……构成 |
| consist of | 由……构成 |
| be composed of | 由……构成 |
| pay streak | 生产层 |
| horizontal drilling | 水平钻井 |
| hydraulic fracturing | 水力压裂 |
| fracturing fluids | 压裂液 |

## Language Focus

1. Existing three to six kilometers underground, they are typically found at much greater subsurface depths than conventional gas reservoirs.

（该句和正文中上句可一起译为:含气页岩是富含有机质的地层,总有机碳重量占其成分的25%,一般埋藏于地下3~6km处,远深于常规气藏。）

本句中,Existing three to six kilometers underground 为现在分词短语做状语,其逻辑主语是句子的主语 they,翻译时可省略。

2. Natural gas is considered "dry" when it is almost pure methane, having had most of the other commonly associated hydrocarbons removed.

（参考译文：把大部分通常伴生的碳氢化合物去除后，天然气几乎就是纯净的甲烷了，这种加工后的天然气称为"干气"。）

本句中，when 引导时间状语从句。having 引导的分词短语做状语，表示先一步的动作。

3. This not only promotes increased reservoir contact, but also provides access to gas reservoirs that were previously difficult or impossible to reach, such as those that run under heavily populated areas or geologic features like lakes.

（参考译文：这不仅有利于接触油藏，而且还能进入之前难以到达的油气藏，例如处于人口稠密的居住区下面或者湖泊下面的气藏。）

本句主句中，not only... but also 关联词语连接两个谓语；reservoirs 后和 such as those 后各跟有一个 that 引导的定语从句，第一个 that 指代其先行词 gas reservoirs，第二个 that 指代其先行词 those，而 those 又指代前面的 gas reservoirs。

## Reinforced Learning

### Ⅰ. Answer the following questions for a comprehension of the text.

1. How much natural gas do we consume annually?

2. What are the characteristics of the shale that distinguish itself from other conventional reservoir?

3. What results in the low permeability of shale?

4. Describe the composition of natural gas.

5. Describe the impacts that theadvances in exploration, drilling, production and horizontal drilling technologies have had on the exploitation of natural gas.

### Ⅱ. Multiple choice: choose the correct one from the alternative answers to give the exact meaning of the words.

1. Estimated to exceed 18,775 trillion cubic feet (Tcf), the resource potential of gas from shales underscores exceeds that of all total world conventional resources combined.

    A. surpasses    B. declines    C. decreases    D. accumulates

2. Existing three to six kilometers underground, they are typically found at much greater subsurface depths than conventional gas reservoirs.

    A. superpositions  B. rocks    C. accumulations  D. wells

3. Further, they are <u>horizontally</u> oriented underground.

 A. slantly   B. flatly   C. levelly   D. vertically

4. Sufficiently brittle such that they can be fractured and rigid enough to maintain open <u>fractures</u>.

 A. breaks   B. scars   C. damages   D. loses

5. Such <u>formations</u> are therefore viewed as both source rocks and reservoirs for natural gas and are often called gas shales.

 A. organizations B. classifications C. separations  D. independences

6. Table 2 −2 highlights such compositional <u>variability</u>.

 A. differences  B. variousness  C. advantages  D. disadvantages

7. Considerable advances in <u>exploration</u>, drilling, and production technologies have made accessing and recovering gas from low-permeability shales more economically viable over the past two decades.

 A. borehole   B. examination  C. surveying  D. prospecting

8. Promoting gas flow from low-permeability shales has been further facilitated by advances in reservoir <u>stimulation</u> techniques.

 A. increase   B. refuse   C. decline   D. strengthening

**Ⅲ. Multiple choice: read the four suggested translations and choose the best answer.**

1. These characteristics of shale resources add significant <u>complexity</u> and cost to gas exploration and production and distinguish shale from conventional natural gas resources.

 A. 变化性  B. 复杂性  C. 多发性  D. 易燃性

2. The naturally tabular clay grains tend to lie flat as the sediments <u>accumulate</u>.

 A. 积累   B. 扩大   C. 减少   D. 消失

3. The following table shows the typical compositional <u>ranges</u> of shale gas produced in the United States.

 A. 方式   B. 变化   C. 区别   D. 类别

4. Advances in <u>horizontal drilling</u> have improved drilling efficiency and gas production rates.

 A. 水平钻井 B. 水中钻井 C. 海中钻井  D. 地平线钻井

5. The process is essentially a controlled fracturing of shale rock to create

conductive paths for gas to flow to the wellbore.

A. 建设的      B. 导流的      C. 快捷的      D. 复杂的

**Ⅳ. Put the following sentences into Chinese.**

1. Even if only 25 percent of this total resource base becomes technically recoverable, shale gas alone can meet global natural gas demand for over 40 years at current consumption levels.

2. Over millions of years, the organic material trapped in shale becomes compressed and heated, forming hydrocarbons including natural gas.

3. With years of deposition, these sediments become highly compacted and solidify into thinly layered shalerock.

4. The absence or presence of geologic features such as faults is another factor which distinguishes shales from one another.

5. The process is essentially a controlled fracturing of shale rock to create conductive paths for gas to flow to the wellbore.

**Ⅴ. Put the following paragraphs into Chinese.**

1. Gas shales are organic-rich formations with total organic carbon accounting for up to 25 percent of their composition by weight. Existing three to six kilometers underground, they are typically found at much greater subsurface depths than conventional gas reservoirs. Shales have low permeability matrices and act as both the source and reservoir rock for gas. Further, they are horizontally oriented underground. These characteristics of shale resources add significant complexity and cost to gas exploration and production and distinguish shale from conventional natural gas resources.

2. Considerable advances in exploration, drilling, and production technologies have made accessing and recovering gas from low-permeability shales more economically viable over the past two decades. Advances in horizontal drilling have improved drilling efficiency and gas production rates. These technologies enable wells to reach and accurately follow long, narrow, horizontal pay streaks. This not only promotes increased reservoir contact, but also provides access to gas reservoirs that were previously difficult or impossible to reach, such as those that run under heavily populated areas or geologic features like lakes.

# 1.3　Tar Sands

## ⊞ Guidance to Reading

*Oil sands, tar sands or, more technically, bituminous sands, are a type of unconventional petroleum deposit. Much of the world's oil is in the form of tar sands, although it is not all recoverable. Oil sands reserves have only recently been considered to be part of the world's oil reserves, as higher oil prices and new technology enable profitable extraction and processing. However, tar sands oil is the worst type of oil for the climate, producing three times the greenhouse gas emissions of conventionally produced oil because of the energy required to extract and process tar sands oil.*

## ⊞ Text

**Tar sands** (also referred to as **oil sands**) are a combination of clay, sand, water, and **bitumen**, a heavy black **viscous** oil, as shown in Figure 1.2. Tar sands can be mined and processed to extract the oil-rich bitumen, which is then refined into oil. The bitumen in tar sands cannot be pumped from the ground in its natural state; instead tar sand deposits are mined, usually using **strip mining** or **open pit techniques**, or the oil is extracted by underground heating with additional **upgrading**.

Tar sands are mined and processed to generate oil similar to oil pumped from conventional oil wells, but extracting oil from tar sands is more complex than conventional oil recovery, as shown in Figure 1.3. Oil sands recovery processes include **extraction** and **separation** systems to separate the bitumen from the clay, sand, and water that make up the tar sands. In order to extract the bitumen from the oil sands, the ore is mixed with warm water to create a **slurry**. This slurry is fed into a processing unit where the bitumen is separated from the water and sand mixture. Bitumen also requires additional upgrading before it can be refined. Because it is so viscous (thick), it also requires dilution with lighter hydrocarbons to make it transportable by pipelines. The extracted bitumen is **diluted** with a special **solvent** and then sent via pipeline to an upgrading facility, where it is transformed into a wide range of premium low-sulfur and low-viscosity synthetic crude oils.

**Figure 1. 2   Tar Sands**         **Figure 1. 3   The Tar Sands Production**

Currently, oil is not produced from tar sands on a significant commercial level in the United States; in fact, only Canada has a large-scale commercial tar sands industry, though a small amount of oil from tar sands is produced commercially in Venezuela.

The Canadian tar sands industry is centered in Alberta, and more than one million barrels of synthetic oil are produced from these resources per day. Currently, tar sands represent about 40% ofCanada's oil production, and output is expanding rapidly. Approximately 20% of U. S. crude oil and products come from Canada, and a substantial portion of this amount comes from tar sands. The tar sands are extracted both by mining and **in situ recovery methods**. Canadian tar sands are different than U. S. tar sands in that Canadian tar sands are water wetted, while U. S tar sands are hydrocarbon wetted. As a result of this difference, extraction techniques for the tar sands in Utah will be different than for those in Alberta.

Recently, prices for crude oil have again risen to levels that may make tarsands-based oil production in theUnited States commercially attractive, and both government and industry are interested in pursuing the development of tar sands oil resources as an alternative to conventional oil.

### Tar Sands Extraction and Processing

Tar sands deposits near the surface can be recovered by open pit mining techniques. New methods introduced in the 1990s considerably improved the efficiency of tar sands mining, thus reducing the cost. These systems use large hydraulic and electrically powered **shovels** to dig up tar sands and load them into enormous trucks that can carry up to 320 tons of tar sands per load, as shown in Figure 1. 4.

After mining, the tar sands are transported to an extraction plant, where a hot water process separates the bitumen from sand, water, and minerals. The separation takes place in separation cells, as shown in Figure 1. 5. Hot water is added to the sand, and the resulting slurry is piped to the extraction plant where it is **agitated**. The combination of hot water and agitation releases bitumen from the oil sand, and causes tiny air bubbles to attach to the bitumen droplets, that float to the top of the separation vessel, where the bitumen can be **skimmed off**. Further processing removes residual water and solids. The bitumen is then transported and eventually upgraded into **synthetic crude oil**.

Figure 1. 4   Tar Sands Open
Pit Mining, Alberta, Canada

Figure 1. 5   Tar Sands Extraction
Separation Cell, Alberta, Canada

About two tons of tar sands are required to produce one barrel of oil. Roughly 75% of the bitumen can be recovered from sand. After oil extraction, the **spent** sand and other materials are then returned to the mine, which is eventually **reclaimed**.

In-situ production methods are used on bitumen deposits buried too deep for mining to be economically recovered. These techniques include **steam injection**, solvent injection, and **firefloods**, in which oxygen is injected and part of the resource burned to provide heat. So far steam injection has been the favoured method. Some of these extraction methods require large amounts of both water and energy (for heating and pumping).

Both mining and processing of tar sands involve a variety of environmental impacts, such as global warming and greenhouse gas emissions, disturbance of mined land; impacts on wildlife and air and water quality. The development of a commercial tar sands industry in the U. S. would also have significant social and economic impacts on local communities. Of special concern in the relatively arid western United States is the large amount of water required for tar sands processing; currently, tar sands extraction and processing require several barrels of water for each barrel of oil produced, though some of the water can be recycled.

The Tar Sands "Gigaproject" is the largest industrial project in human history and likely also the most destructive. The tar sands mining procedure releases at least three times the $CO_2$ emissions as regular oil production and is slated to become the single largest industrial contributor in North America to Climate Change.

The tar sands are already slated to be the cause of up to thesecond fastest rate of **deforestation** on the planet behind the Amazon Rainforest Basin. Currently approved projects will see 3 million barrels of tar sands mock crude produced daily by 2018; for each barrel of oil up to as high as five barrels of water are used.

Human health in many communities has seriously **taken a turn for the worse** with many causes alleged to be from tar sands production. Tar sands production has led to many serious social issues throughout Alberta, from housing crises to the vast expansion of temporary foreign worker programs that racialize and exploit so-called non-citizens. Infrastructure from pipelines to refineries to super tanker oil traffic on the seas crosses the continent in all directions to all three major oceans and the Gulf of Mexico.

## 🔲 Words and Expressions

| | |
|---|---|
| bitumen | 沥青,柏油 |
| viscous | 黏性的,黏的;黏滞的 |
| upgrading | 浓缩;加工 |
| extraction | 提取,萃取 |
| separation | 分选(矿);分开 |
| slurry | 泥浆,悬浮液 |
| dilute | 稀释的,淡的;稀释,冲淡 |
| solvent | 溶剂;解决方法 |
| shovel | 铲(挖斗机,一铲的量) |
| agitate | 搅动,煽动 |
| spent | 废的(用过的) |
| reclaim | 开垦,收回,回收再利用 |

| firefloods | 火驱(法) |
| deforestation | 采伐森林,森林砍伐 |

## 🔲 Phrases and Expressions

| tar sand/oil sand | 沥青砂;含油砂,焦油砂 |
| strip mining | 露天开采 |
| open pit technology | 露天开采技术 |
| in situ | 在原地(就地,在现场) |
| recovery method | 采油方法 |
| skim off | 撇去 |
| synthetic crude oil | 合成原油 |
| steam injection | 注蒸汽 |
| taken a turn for the worse | 开始恶化 |

## 🔲 Language Focus

1. Tar sands can be mined and processed to extract the oil-rich bitumen, which is then refined into oil.

（参考译文：通过开采和加工焦油砂,可获取富含石油的沥青质,然后再将沥青质炼制成石油。）

本句中,which 引导非限制性定语从句,修饰 oil-rich bitumen。

2. After oil extraction, the spent sand and other materials are then returned to the mine, which is eventually reclaimed.

（参考译文：采油结束后,废弃油砂及其他物质都要运回矿场,使矿场恢复再生。）

本句中,which 引导非限制性定语从句修饰 mine。

3. New methods introduced in the 1990s considerably improved the efficiency of tar sands mining, thus reducing the cost.

（参考译文：20世纪90年代引进的新方法大大提高了焦油砂的开采效率,从而降低了成本。）

本句为简单句,主谓结构是 New methods improved...。主语后的 introduced in the 1990s 是过去分词短语作其定语。thus reducing the cost 是现在分词短语作状语表示结果。

4. Of special concern in the relatively arid western United States is the large amount of water required for tar sands processing.

（参考译文：美国西部地区相对干旱，油砂加工需要耗费大量的水，因此引起了人们的特别关注。）

本句是倒装句，正常语序是：The large amount of water required for tar sands processing is of special concern in the relatively arid western United States. 其中 required for tar sands processing 是过去分词短语做 water 的后置定语。

## Reinforced Learning

**I. Answer the following questions for a comprehension of the text.**

1. What are tar sands?

2. What is special about oil sands extraction compared with conventional oil recovery?

3. What do the in-situ production methods include?

4. What kinds of environmental impacts do the mining and processing of tar sands bring?

5. What is the difference between Canadian tar sands and US tar sands?

**II. Multiple choice：choose the correct one from the alternative answers to give the exact meaning of the words.**

1. Oil sands recovery processes include <u>extraction</u> and separation systems to separate the bitumen from the clay, sand, and water that make up the tar sands.

　A. abstracting　　B. bringing　　C. joining　　D. mixing

2. The combination of hot water and agitation releases bitumen from the oil sand, and causes tiny air bubbles to attach to the bitumen droplets that float to the top of the <u>separation</u> vessel, where the bitumen can be skimmed off.

　A. sequence　　B. combination　C. segregation　D foundation

3. Bitumen also requires additional <u>upgrading</u> before it can be refined.

　A. getting　　B. mixing　　C. composing　D. processing

4. These systems use large hydraulic and electrically powered shovels to dig up tar sands and load them into <u>enormous</u> trucks that can carry up to 320 tons of tar sands per load.

　A. giant　　B. diminutive　C. small　　D. pygmy

5. The tar sands mining procedu <u>rereleases</u> at least three times the $CO_2$ emissions as regular oil production and is slated to become the single largest industrial contributor in North America to Climate Change.

A. absorbs　　　　B. expels　　　　C. captures　　　　D. takes in

**III. Multiple choice：read the four suggested translations and choose the best answer.**

1. Tar sands are mined and processed to generate oil similar to oil pumped from conventional oil wells，but extracting oil from tar sands is more complex than conventional oil recovery.

　　A. 保守的　　　　B. 常规的　　　　C. 合约的　　　　D，传统的

2. Currently，oil is not produced from tar sands on a significant commercial level in the United States

　　A. 商业的　　　　B. 工业的　　　　C. 工艺的　　　　D. 经济的

3. After mining，the tar sands are transported to an extraction plant，where a hot water process separates the bitumen from sand，water，and minerals.

　　A. 岩石　　　　B. 提取物　　　　C. 矿物质　　　　D. 分子

4. These techniques include steam injection，solvent injection，and fire-floods，in which oxygen is injected and part of the resource burned to provide heat.

　　A. 放入　　　　B. 注射　　　　C. 提取　　　　D. 萃取

5. Infrastructure from pipelines to refineries to super tanker oil traffic on the seas crosses the continent in all directions to all three major oceans and the Gulf of Mexico.

　　A. 建筑物　　　　B. 体育器材　　　C. 工厂　　　　D. 基础设施

**IV. Put the following sentences into Chinese.**

1. Tar sands can be mined and processed to extract the oil-rich bitumen，which is then refined into oil.

2. Tar sands are mined and processed to generate oil similar to oil pumped from conventional oil wells，but extracting oil from tar sands is more complex than conventional oil recovery.

3. After oil extraction，the spent sand and other materials are then returned to the mine，which is eventually reclaimed.

4. These techniques include steam injection，solvent injection，and fire-floods，in which oxygen is injected and part of the resource burned to provide heat.

5. Human health in many communities has seriously taken a turn for the

worse with many causes alleged to be from tar sands production.

### V. Put the following paragraphs into Chinese.

1. Tar sands deposits near the surface can be recovered by open pit mining techniques. New methods introduced in the 1990s considerably improved the efficiency of tar sands mining, thus reducing the cost. These systems use large hydraulic and electrically powered shovels to dig up tar sands and load them into enormous trucks that can carry up to 320 tons of tar sands per load.

2. The tar sands are already slated to be the cause of up to the second fastest rate of deforestation on the planet behind theAmazon Rainforest Basin. Currently approved projects will see 3 million barrels of tar sands mock crude produced daily by 2018; for each barrel of oil up to as high as five barrels of water are used.

## 1.4　Natural Gas Liquids

### 🔲 Guidance to Reading

*Natural gas liquids (NGLs) are also hydrocarbons which are separated from the gas state in the form of liquids. NGLs are valuable as separate products and it is therefore profitable to remove them from the natural gas. The liquids are first extracted from the natural gas and later separated into different components. There are many uses for NGLs, spanning nearly all sectors of the economy. NGLs are used as inputs for petrochemical plants, burned for cooking, and blended into vehicle fuel. Higher crude oil prices have contributed to increased NGL prices and, in turn, provided incentives to drill in liquids-rich resources with significant NGL content.*

### 🔲 Text

Natural Gas, the source ofNatural Gas Liquids is a natural occurring mixture of gaseous hydrocarbons found in the ground or obtained from specially driven wells. The composition of natural gas varies in different parts of the world. Its chief component, **methane**, usually makes up from 80% to 95% of its composition. The balance is composed of varying amounts of **ethane**, **propane**, **butane**, and other hydrocarbon compounds.

NGLs are produced via **refrigeration** and **distillation** processes that take place in gas plants and refineries. They are considered "by-products" of the oil and gas industry. Gas plants extract NGLs for profit and/or to ensure production of **pipeline quality** natural gas.

Natural Gas Liquids are the raw, associated gases and liquids that come up along with oil and natural gas from the well. NGLs are very important—vital even—now for regular, dry gas (methane) producers, as they are separated and sold as more expensive products like ethane, propane, butane and **condensate**.

US producers have tried to offset the impact of depressed gas prices by shifting their drilling from so-called "**dry gas**" **wells**, which produce only gas, to "**wet wells**", which produce a mix of gas and more expensive oil liquids for the **petrochemical** industry.

The conditions supportive of unconventional gas production also support increased output of natural gas liquids (NGLs), extracted from liquids-rich shale gas, as well as light tight oil. This oil is **analogous** in many ways to shale gas, both in terms of its origins—it is oil that has not migrated, or at least not migrated far, from the (shale) source rock—and in terms of the production techniques required to exploit it. Light tight oil is being produced from many of the same basins as unconventional gas in the United States, and, in a price environment combining high oil prices and very low prices for natural gas, there is a strong economic **incentive** to target plays with higher liquids content.

United States' NGL production from shales is increasing rapidly and up to 1 mb/d of new capacity is expected to be added by 2020. The growth in NGL production is creating new opportunities for the petrochemical industry, but action will be required to remove pipeline **bottlenecks** and provide additional **fractionation** and storage facilities if the benefits are to be fully realized. The growth in global production of NGLs from shale formations and **light tight oil** in the period to 2020, predominantly in North America, makes up almost half the **incremental** growth in oil supply over this period.

Production outside North America of NGLs from shale and of light tight oil is unlikely to make a large contribution to global liquids production before 2020 as much **evaluation** work still needs to be done. However, the Neuquén basin in Argentina shows promise, YPF announcing potential resources of 7 billion barrels, while the extension of the Eagle Ford shale into Mexico is also a focus of attention. Our projections for light tight oil production outside North America remain small even beyond 2020, as we have yet to see sufficient progress in confirming resources, so there is some **upside potential**.

It should be noted, however that on the basis of current knowledge, light tight oil resources are expected to be of less consequence than shale gas resources: whereas the estimated shale gas resources in the United States represent at least 5 years of 2010 domestic gas demand, the known light tight oil resources make up no more than four years of domestic oil demand. This is why we currently **predict** light tight oil production in theUnited States to peak in the 2020s.

The liquids content of shale gas plays is an important consideration in their economic **viability** as NGLs are easily transported to world markets, while market opportunities for gas are often only local, at prices that may not be **aligned** to international prices for reasons of policy or **infrastructure**. However there is always a degree of uncertainty about the extent of liquids content until new shales have been drilled and tested.

Future growth inUS natural gas liquids (NGLs) output will depend on drillers' ability to use **hydraulic** fracturing to **complete wells** in **shale formations**. But the technique remains increasingly **controversial**. Hydraulic fracturing of wells involves injecting a high-pressure mix of water, sand and chemicals to help break and hold open dense rock formations, allowing crude, gas and NGLs to flow to the surface. But environmental groups say the process uses too much water, threatens underground water quality, pollutes the air and even causes minor earthquakes. The US **Environmental Protection Agency** (EPA) is carrying out a study on its impact on drinking water and plans a national drilling wastewater treatment standard. Pennsylvania and Texas have passed **fracturing fluid** disclosure laws in the past two years.

NGL output in the first quarter of this yearaccounted for a record of almost 30 percent of the US total oil production. The shale oil boom is causing a steep price drop in Natural Gas Liquids (NGLs) in North America, hurting gas producers. Natural gas liquids **make the money** for many shale gas producers. And that could, at times, **cut into** shale gas production. A **Btu** of oil is selling for more than 20 times the price of a Btu of natural gas in U. S. markets. The liquids accompanying shale gas production growth could drive down ethane prices. Good for customers, bad for producers. When removed from the gas stream, ethane is a feedstock for ethylene, used to make plastics.

Because revenue from gas liquids is so important to shale producers, especially when natural gas prices are low, a drop in the ethane price could push some drillers to **back off** production until the ethane supply falls more **in line with** demand and prices rise again.

To illustrate just how important the revenue from gas liquids can be for a shale gas producer, ran through some numbers:

A typical shale well needs $5 to $6 per thousand cubic feet of output to make money.

Methane has been priced at about $4 most of the past year.

But the added value of the liquids makes the entire flow from the well worth about $7 to $8 per mcf.

Many of the experts discussed building up the market for natural gas and its liquids to ensure that shale and other unconventional gas such as tight sands-and **Arctic** gas, too-will find buyers willing to pay a decent price.

## 🔲 Words and Expressions

| | |
|---|---|
| methane | 甲烷 |
| ethane | 乙烷 |
| propane | 丙烷 |
| butane | 丁烷 |
| refrigeration | 制冷;冷却 |
| distillation | 蒸馏 |
| condensate | 凝析油 |
| offset | 抵消,补偿,弥补 |
| petrochemical | 石油化学产品;石化的 |
| analogous | 类似的;可比拟的 |
| incentive | 动机;刺激 |
| bottleneck | 瓶颈;障碍物 |
| fractionation | 分别;分馏法 |
| incremental | 增加的,增值的 |
| evaluation | 评价;评估;估价;求值 |
| viability | 生存能力,发育能力;可行性 |
| align | 使联盟;匹配;排成一行 |

| | |
|---|---|
| infrastructure | 基础设施;公共建设 |
| hydraulic | 液压的;水力的 |
| controversial | 有争议的;有争论的 |
| Btu | 英热(英国热量单位) |
| Arctic | 北极圈;北极的 |

## ⌐┘ Phrases and Expressions

| | |
|---|---|
| Natural Gas Liquids | 天然气凝析液 |
| pipeline quality | (符合)管输质量(要求的) |
| dry gas well | 干气井 |
| wet well | 湿气井;出液体的井 |
| light tight oil | 轻致密油 |
| upside potential | 上涨潜力,上升空间 |
| complete wells | 完井 |
| shale formation | 页岩地层 |
| Environmental Protection Agency | (美)环境保护局 |
| fracturing fluid | 压裂液 |
| make money | 赚钱 |
| cut into | 减少 |
| back off | 倒开 |
| in line with | 按照;与……一致 |

## ⌐┘ Language Focus

1. US producers have tried to offset the impact of depressed gas prices by shifting their drilling from so-called "dry gas" wells, which produce only gas, to "wet wells", which produce a mix of gas and more expensive oil liquids for the petrochemical industry.

(参考译文:鉴于天然气价格低迷,美国生产商不再钻所谓只产气的"干气"井,而是转向钻"湿气井","湿气井"可产出气液混合体,其中液体更贵重,适用于石油化工业。)

本句中,介词 by 后用了动名词短语 shifting...to...,意为"从……转向……"。两个 which 均引导非限制性定语从句,第一个 which 修饰"dry gas" wells,第二个 which 修饰"wet wells"。

2. It should be noted, however that on the basis of current knowledge, light

tight oil resources are expected to be of less consequence than shale gas resources: whereas the estimated shale gas resources in the United States represent at least 5 years of 2010 domestic gas demand, the known light tight oil resources make up no more than four years of domestic oil demand.

（参考译文：然而，应该注意的是，从目前掌握的资料来看，轻致密油资源没有页岩气资源多：按照美国2010年国内天然气需求量来估计，美国页岩气资源至少能用5年；而已探明的轻致密油资源最多也只能满足国内4年的石油需求。）

本句中，It 为形式主语，代替 that 引导的从句，that 从句内又包含一个 than 引导的比较状语从句；该句后面又出现一个说明性复合句，the known light tight oil resources make up… 为主句，whereas 引导状语从句，表示对比。

3. The liquids content of shale gas plays is an important consideration in their economic viability as NGLs are easily transported to world markets, while market opportunities for gas are often only local, at prices that may not be aligned to international prices for reasons of policy or infrastructure.

（参考译文：凝析液含量是决定页岩气生产经济可行性的一个重要因素，因为天然气凝析液易于运至国际市场；而天然气一般只在当地市场交易，由于政策和基础设施等原因其价格无法与国际价格接轨。）

本句用 while 引导一个让步状语从句，来对比页岩气和页岩气凝析液。主句本身也是复合句，含有一个 as 引导的原因状语从句；while 从句中又含有一个 that 引导的定语从句，修饰其先行词 prices。

4. The liquids accompanying shale gas production growth could drive down ethane prices. Good for customers, bad for producers.

（参考译文：随着页岩气产量的增长，天然气凝析液可能会导致乙烷价格下降。这对顾客是好事，而对生产商却是坏事。）

本部分中，前一句的主谓结构是 The liquids could drive down ethane prices，主语后的 accompanying… 为现在分词短语做其后置定语。后一句 Good for customers, bad for producers. 为省略句，good 和 bad 前均省略了 it is。

## Reinforced Learning

### I. Answer the following questions for a comprehension of the text.

1. What's the chief component of NGL?
2. Give a definition of NGLs.
3. Describe what's hydraulic fracturing.

4. What problems may hydraulic fracturing cause?

5. Why do people currently project light tight oil production in the United States to peak in the 2020s?

**II. Multiple choice: choose the correct one from the alternative answers to give the exact meaning of the words.**

1. Its chief component, methane, usually makes up from 80% to 95% of its composition.

    A. thesis      B. framework     C. compotation     D. constitution

2. Gas plants extract NGLs for profit and/or to ensure production of pipeline quality natural gas.

    A. refine      B. remove     C. abstract     D. excerpt

3. US producers have tried to offset the impact of depressed gas prices by shifting their drilling from so-called "dry gas" wells, which produce only gas, to "wet wells", which produce a mix of gas and more expensive oil liquids for the petrochemical industry.

    A. eliminate    B. counteract     C. conduct     D. vanish

4. NGLs are very important—vital even—now for regular, dry gas (methane) producers, as they are separated and sold as more expensive products like ethane, propane, butane and condensate.

    A. vicious     B. viral     C. significant     D. immortal

5. Production outside North America of NGLs from shale and of light tight oil is unlikely to make a large contribution to global liquids production before 2020 as much evaluation work still needs to be done.

    A. estimation    B. projection     C. development     D. assessment

6. This oil is analogous in many ways to shale gas, both in terms of its origins—it is oil that has not migrated, or at least not migrated far, from the (shale) source rock—and in terms of the production techniques required to exploit it.

    A. anonymous  B. identical     C. similar     D. various

7. Future growth inUS natural gas liquids (NGLs) output will depend on drillers' ability to use hydraulic fracturing to complete wells in shale formations. But the technique remains increasingly controversial.

    A. conventional              B. debatable

    C. convivial               D. unquestionable

8. But environmental groups say the process uses too much water, threatens underground water quality, pollutes the air and even causes minor earthquakes.

　　A. major　　　B. slight　　　C. many　　　　D. minimum

9. Many of the experts discussed building up the market for natural gas and its liquids to ensure that shale and other unconventional gas such as tight sands- and Arctic gas, too-will find buyers willing to pay a decent price.

　　A. reasonable　B. recent　　　C. honorable　　D. noble

10. The liquids accompanying shale gas production growth could drive down ethane prices.

　　A. along with　B. after　　　C. keeping　　　D. ahead of

**III. Multiple choice: read the four suggested translations and choose the best answer.**

1. This is why we currently predict light tight oil production in the United States to peak in the 2020s.

　　A. 发射　　　　B. 预测　　　C. 设计　　　D. 计划

2. The growth in global production of NGLs from shale formations and light tight oil in the period to 2020, predominantly in North America, makes up almost half the incremental growth in oil supply over this period.

　　A. 卓越地　　　B. 优秀地　　C. 客观地　　　D. 主要地

3. Our projections for light tight oil production outsideNorth America remain small even beyond 2020, as we have yet to see sufficient progress in confirming resources, so there is some upside potential.

　　A. 成熟的　　　B. 不足的　　C. 足够的　　　D. 多余的

4. Hydraulic fracturing of wells involves injecting a high-pressure mix of water, sand and chemicals to help break and hold open dense rock formations, allowing crude, gas and NGLs to flow to the surface.

　　A. 打针　　　　B. 投掷　　　C. 注入　　　D. 注射

5. Pennsylvania and Texas have passed fracturing fluid disclosure laws in the past two years.

　　A. 披露　　　　B. 泄露　　　C. 揭发　　　D. 揭示

**IV. Put the following sentences into Chinese.**

1. NGL output in the first quarter of this year accounted for a record of almost 30 per cent of theUS total oil production.

2. The conditions supportive of unconventional gas production also support increased output of natural gas liquids (NGLs), extracted from liquids-rich shale gas, as well as light tight oil.

3. The growth in global production of NGLs from shale formations and light tight oil in the period to 2020, predominantly inNorth America, makes up almost half the incremental growth in oil supply over this period.

4. The shale oil boom is causing a steep price drop in Natural Gas Liquids (NGLs) inNorth America, hurting gas producers.

5. The liquids accompanying shale gas production growth could drive down ethane prices.

**V. Put the following paragraphs into Chinese.**

1. NGLs are produced viarefrigeration and distillation processes that take place in gas plants and refineries. They are considered "by-products" of the oil and gas industry. Gas plants extract NGLs for profit and/or to ensure production of pipeline quality natural gas.

2. The liquids content of shale gas plays is an important consideration in their economicviability as NGLs are easily transported to world markets, while market opportunities for gas are often only local, at prices that may not be aligned to international prices for reasons of policy or infrastructure. However there is always a degree of uncertainty about the extent of liquids content until new shales have been drilled and tested.

# Chapter 2    Oil and Gas Industry

## 2. 1    Oil and Gas Industry Structure

### 🔲 Guidance to Reading

*Unconventionaloil and gas industry can also be classified into three major* **segments**: *upstream, midstream, and downstream. The upstream segment is most involved in the development of unconventional oil and gas, though the extraction methods required are different than conventional oil and gas recovery, while mid-and downstream activities and processes for are the same as those for conventional oil and gas. As a result, the focus of this chapter is on the upstream sector. The upstream segment of the industry seek and acquire rights to potential oil and gas fields and conduct activities related to the extraction of unconventional hydrocarbons, with companies usually segmented into two categories: gas E&P companies and oilfield services companies. Therefore, the exploration and production industry trends as well as the oilfield services industry trends are also given an eye to.*

### 🔲 Text

The oil and gas industry can be classified into three major segments: upstream, midstream, and downstream.

- Upstream companies are involved in the exploration, development, and production of crude or natural gas, also called exploration and production, or E&P.

- Midstream operations create **linkage** between producing areas, which are typically remote, and the consumer base. Midstream activities include the processing, storage, marketing, and bulk transportation of petroleum commodities.

- Downstream industry participants include oil refineries, petrochemical plants, natural gas distribution companies, distributors of petroleum products, and retail outlets.

These companies provide thousands of petroleum products to consumers such as gasoline, **diesel**, natural gas, heating oil, **asphalt**, **lubricants**, plastics, fertilizers, **antifreeze**, **pesticides**, and propane.

The upstream segment is most involved in the development of shale gas while mid-and downstream activities and processes for shale gas are the same as those for conventional natural gas. Companies that participate in the upstream segment of the industry seek and acquire rights to potential oil and gas fields and conduct activities related to the recovery and production of hydrocarbons. These companies can be further segmented into two categories: gas E&P companies and oilfield services (OFS) companies.

E&P companies search for potential oil and gas fields and drill and operate wells that recover oil and gas from underground formations. Company revenues are substantially derived from sales of petroleum produced at the wellhead. In theUnited States, two types of E&P companies produce and sell gas to pipelines and other midstream participants: international and independent oil companies.

International oil companies (IOCs) are non-state owned, publicly traded large-cap energy companies such as BP, ExxonMobil, and Royal Dutch Shell. These companies are frequently forward integrated into mid-and downstream activities.

Independent companies, or "independents", are non-integrated, much smaller companies compared to players like BP and ExxonMobil and derive revenues substantially from wellhead production of gas. These companies are much less frequently forward integrated and where they are, activities tend to be limited to marketing and midstream operations for their own produced gas. Independents can range in size from multi-billion dollar publicly traded companies to small **proprietorships**. Examples of independent gas companies active in US shale gas development include Chesapeake Energy, Southwestern Energy, and Petrohawk Energy.

In various countries outside of theUS, gas is also produced by national oil companies (NOCs). These are state-controlled oil and gas companies fully or majority owned by national governments. NOCs control a substantial portion of the world's oil and gas reserves. Russia's Gazprom, Petrobras in Brazil, and Petro China are three examples of major NOCs worldwide.

Finally, oilfield services companies support the upstream petroleum sector. These companies typically do not derive revenues from their own petroleum production, but instead provide technology, **consulting**, project management, well services and maintenance, and information solutions to the oil and gas exploration and production industry. Product offerings from businesses in this segment include various scientific and technical services such as **seismic**evaluation and analysis, drilling equipment and services, engineering services, and supplies including drill bits and fracturing fluids. OFS companies can range in size and scope from small proprietorships which offer specialized products and services to multi-billion dollar integrated service companies. The three largest OFS companies in the world are Schlumberger, Halliburton and Baker Hughes.

### *Occupations in Upstream Oil and Gas*

There are a wide range of jobs in the upstream segment of the oil and gas industry. E&P and OFS companies require resources to fill hundreds of various entry level, technical, professional, and management positions. Professional occupations account for a quarter of jobs in E&P companies. Nearly 80 percent of these are related to architecture, engineering, and physical sciences. Typical professionals include engineers, **geologists**, geophysicists, chemical and environmental scientists, draftsmen, cartographers, and support technicians. Most of these jobs require a four-year degree and many may require advanced degrees. Technicians in the industry typically require a two to four year degree. Specific roles include:

- Geophysicists to determine likely locations of hydrocarbon reservoirs using **gravitational**, magnetic, electrical, and seismic methods.

- Geologists to survey and assess the gas producing potential of a reservoir.

- Geographic Information Specialists to create and maintain databases with information related to gas well site locations, pipelines, compressor stations, roads, and other facilities.

- Environmental Technicians to develop and implement environmental plans and perform soil and water testing analysis to ensure compliance with environmental law and regulations.

- Field engineers to supervise drilling, completion, and well workover operations.

Non-management functions related to construction and extraction are the third largest category of jobs in E&P companies and account for over 16 percent of the segment's employment. The vast majority of these occupations are related to activities that occur during ongoing hydrocarbon extraction operations. Examples of jobs directly related to extraction include production foremen, well tenders, and compressor operators. E&P companies also **leverage excavation** contractors and

**timber** cutters to clear land and build roads, well sites, tanks, compressor stations, pipelines, and power lines.

In several aspects, oilfield services companies employ similar resources to gas extraction companies. However, a difference in employment mix exists between the two types of companies. In E&P companies, professional occupations account for the largest number of jobs. In contrast, production occupations account for the largest number of jobs in OFS companies. In fact, over 57 percent of total OFS jobs are well site labor jobs.

Two job categories related to drilling account for over 20 percent of OFS employment. These include **roustabouts** and rotary drill operators. Roustabouts account for the largest portion, nearly 15 percent, of total jobs in OFS companies. Service unit operators use specialized equipment to perform well stimulation treatments and remove obstructions from the well, accounting for nearly 10 percent of employment in the US drilling and support sector. Many OFS providers also employ research and development engineers and product managers to design and develop new tools and technologies. Further, they employ resources to perform various industry consulting engagements and manage projects.

### Projected Upstream Oil and Gas Employment

Between 2015 and 2020, employment is expected to increase 3.6 percent to nearly 528,000. The growth in industry employment **forecast** by **SBI Energy** is supported by projections made by the **US Energy Information Administration**'s 2010 Annual Energy Outlook. The EIA expects US oil and gas production to increase with consumption demand and also forecasts that crude oil prices will grow from about $78 in 2010 to over $108 per barrel in 2020. This forecast growth indicates that 63,000 new jobs are expected to be created over the next five years and 82,000 over the next 10.

Furthermore, the upstream oil and gas industry faces substantial **attrition** over the coming years and industry participants expect to see 30 to 46 percent of their workforces retire by 2019. The confluence of new jobs created by price improvement and increased drilling and production with massive retirement numbers will create significant employment opportunities over the next five to 10 years. Industry demand will be greatest for qualified professionals and extraction workers with strong technical skills and previous experience in the oil and gas extraction industry.

## Words and Expressions

| | |
|---|---|
| segment | 段;部分;分割 |
| linkage | 连接;结合 |
| diesel | 柴油机;柴油 |
| asphalt | 沥青;柏油 |
| lubricant | 润滑剂;润滑油 |
| antifreeze | 防冻剂 |
| pesticide | 杀虫剂 |
| proprietorship | 所有权 |
| consult | 请教;商议;当顾问 |
| seismic | 地震的;因地震而引起的 |
| geophysics | 地球物理学 |
| gravitational | 重力的,引力的 |
| magnetic | 地磁的;有磁性的;有吸引力的 |
| leverage | 手段,影响力;杠杆作用;利用 |
| excavation | 挖掘;发掘 |
| timber | 木材;木料 |
| roustabout | 甲板工 |
| forecast | 预测,预报;预示 |
| attrition | 摩擦;磨损;消耗 |

## Phrases and Expressions

| | |
|---|---|
| E&P(exploration and production) companies | 勘探与开发公司 |
| Oilfield services (OFS) companies | 油田服务公司 |
| SBI Energy (Specialist In Business Information) | 商业信息专家 |
| US Energy Information Administration (EIA) | 美国能源信息署 |

## Language Focus

1. Midstream operations create linkage between producing areas, which are

typically remote, and the consumer base.

（参考译文：中游业务往往是把比较偏远的产油区和消费区连接起来。）

本句主句结构为：Midstream operations create linkage between… and…。which 引导非限制性定语从句，修饰 producing areas。

2. These companies are much less frequently forward integrated and where they are, activities tend to be limited to marketing and midstream operations for their own produced gas.

（参考译文：这些公司业务极少涉及中下游，即使有的话，也仅限于运输和销售自产天然气。）

本句是一个由 and 连接的并列复合句，and 前面的分句使用了一个比较级，than 及后面的部分省略了，这是国际石油公司和独立石油公司间的比较；and 后面的分句是一个含有 where 状语从句的复合句，where they are 后面省去了 forward integrated。

3. OFS companies can range in size and scope from small proprietorships which offer specialized products and services to multi-billion dollar integrated service companies.

（参考译文：油服公司大小不等，既有提供专门产品和服务的小型独资公司，也有市值数十亿美元，提供一体化服务的大公司。）

本句主句结构为：OFS companies can range in size and scope from… to…。"range from… to…" 表示范围大小；which 引导定语从句，修饰其先行词 small proprietorships。

4. The confluence of new jobs created by price improvement and increased drilling and production with massive retirement numbers will create significant employment opportunities over the next five to 10 years.

（参考译文：由于油价上涨，钻井数量和开采量增加，而同时又有大量从业人员退休，所以很多新的工作岗位将出现，在未来的 5～10 年内，将有大量的就业机会。）

本句只有一个主谓结构，即 The confluence of… will create…，其中 created by 为过去分词短语做后置定语修饰其前面的 new jobs。

## Reinforced Learning

### I. Answer the following questions for a comprehension of the text.

1. Describe the upstream companies' business.

2. What're the two types of E&P companies in theUnited States?

3. How do the oilfield services companies get revenues?

4. What is the third largest category of jobs in E&P companies?

5. Describe the difference in employment mix existing between the oilfield services companies and the gas extraction companies.

**II. Multiple choice: choose the correct one from the alternative answers to give the exact meaning of the words.**

1. The oil and gas industry can be classified into three major segments: upstream, midstream, and downstream.

  A. sections    B. signets    C. secretions    D. cements

2. The upstream segment is most involved in the development of shale gas while mid-and downstream activities and processes for shale gas are the same as those for conventional natural gas.

  A. standard    B. ordinary    C. traditional    D. special

3. These companies can be further segmented into two categories: gas E&P companies and oilfield services (OFS) companies.

  A. deeper    B. better    C. more    D. farther

4. E&P companies search for potential oil and gas fields and drill and operate wells that recover oil and gas from underground formations.

  A. actual    B. latent    C. popular    D. partial

5. Company revenues are substantially derived from sales of petroleum produced at the wellhead.

  A. gained    B. deceived    C. arrived    D. bought

6. In contrast, production occupations account for the largest number of jobs in OFS companies.

  A. consistently    B. in controversial C. on the contrary    D. coincidently

7. Service unit operators use specialized equipment to perform well stimulation treatments and remove obstructions from the well, accounting for nearly 10 percent of employment in the US drilling and support sector.

  A. destructions    B. obstacles    C. constructions    D. structures

8. The growth in industry employment forecast by SBI Energy is supported by projections made by the US Energy Information Administration's 2010 Annual Energy Outlook.

  A. broadcast    B. report    C. conclusion    D. anticipation

9. The confluence of new jobs created by price improvement and increased

drilling and production with massive retirement numbers will create significant employment opportunities over the next five to 10 years.

    A. influence      B. conclusion     C. conflation     D. aggregation

10. These companies are much less frequently forward integrated and where they are, activities tend to be limited to marketing and midstream operations for their own produced gas.

    A. repeatedly    B. functionally    C. constantly    D. regularly

**III. Multiple choice：read the four suggested translations and choose the best answer.**

1. Midstream activities include the processing, storage, marketing, and bulk transportation of petroleum commodities.

    A. 装饰品     B. 商品     C. 日用品     D. 化妆品

2. Field engineers to supervise drilling, completion, and well workover operations.

    A. 监管     B. 控制     C. 查看     D. 照顾

3. E&P companies also leverage excavation contractors and timber cutters to clear land and build roads, well sites, tanks, compressor stations, pipelines, and power lines.

    A. 合同     B. 联系人     C. 承包商     D. 包工头

4. OFS companies can range in size and scope from small proprietorships which offer specialized products and services to multi-billion dollar integrated service companies.

    A. 股份制公司    B. 国有企业    C. 合资企业    D. 独资企业

5. Professional occupations account for a quarter of jobs in E&P companies.

    A. 解释     B. 占     C. 报账     D. 说明

**IV. Put the following sentences into Chinese.**

1. The upstream segment is most involved in the development of shale gas while mid-and downstream activities and processes for shale gas are the same as those for conventional natural gas.

2. E&P company revenues are substantially derived from sales of petroleum produced at the wellhead.

3. NOCs are state-controlled oil and gas companies fully or majority owned

by national governments.

4. Non-management functions related to construction and extraction are the third largest category of jobs in E&P companies and account for over 16 percent of the segment's employment.

5. The growth in industry employment forecast by SBI Energy is supported by projections made by the US Energy Information Administration's 2010 Annual Energy Outlook.

**V. Put the following paragraphs into Chinese.**

1. There are a wide range of jobs in the upstream segment of the oil and gas industry. E&P and OFS companies require resources to fill hundreds of various entry level, technical, professional, and management positions. Professional occupations account for a quarter of jobs in E&P companies. Nearly 80 percent of these are related to architecture, engineering, and physical sciences.

2. Independent companies are non-integrated, much smaller companies compared to players like BP and ExxonMobil and derive revenues substantially from wellhead production of gas. These companies are much less frequently forward integrated and where they are, activities tend to be limited to marketing and midstream operations for their own produced gas. Independents can range in size from multi-billion dollar publicly traded companies to small proprietorships.

# 2.2　Shale Gas Employment Outlook

## Guidance to Reading

*Unconventional oil and gas resources have a great potential with shale gas being dominant, playing a vital role in the global energy framework, and making the key sector in the future of petroleum industry development. However, drilling and **well completion** processes for shale gas wells are significantly more labor intensive than those for conventional gas as they need complex horizontal drilling and **multi-stage hydraulic fracturing** processes. In addition, the economic and employment impacts of shale play development extend well beyond the direct employment of resources in upstream oil and gas. Upstream gas development activities **trigger** a chain of economic and employment impacts through further indirect spending. Therefore, shale gas development will bring about a* *bright future for overall employment.*

## Text

The employment gap caused by retiring workers is a critical strategic challenge for the industry, where the retirement age averages 55 years. In a study released in 2007, the **Interstate Oil and Gas Compact Commission** (**IOGCC**) estimated the average age of US oil and gas industry workers to be 48, with half of the workforce between the ages of 50 and 60. Given the high levels of unemployment currently prevalent in the US, **recruiting** the number of employees needed to backfill retirees should not be **daunting**. However, the challenge lies in replacing the knowledge lost with the attrition of experienced workers in an industry where reaching a level of basic proficiency can require up to three years' field experience.

The upstream oil and gas industry has taken various measures over the past five years to attract talent and address the challenge of the widely publicized "big crew change." Industry organizations have successfully lobbied the federal government to increase internships for geosciences graduates and undergrads. States have provided funding for training programs for lease operators, well service crews, and other field technical resources such as safety engineers. In the meantime, industry associations have conducted massive public relations campaigns designed to **rehabilitate** the industry's poor image.

E&P and OFS companies have strengthened outreach programs to high schools and colleges and have developed mentoring, scholarship, and grant programs to support ongoing education for existing employees. They provide incentives for workers approaching retirement age to stay with the company and leverage **compensation** incentives to attract new talent. Recent petroleum engineering graduates reportedly received starting salaries averaging about $ 75,000 in 2010, making petroleum engineering the most **lucrative** undergraduate degree this year. All of these developments indicate a strong employment outlook for the industry over the next several years.

### Employment in the US Shale Plays

As of July 2010, a total of 1,585 rotary **rigs** were actively drilling in theUnited States, 62 percent of which were drilling for gas. SBI Energy estimates that two-thirds of these gas rigs were drilling for shale gas, indicating that over 40 percent of total US oil and gas drilling activity was devoted to shale gas at the time. Drilling rig activity is a strong indicator of oil and gas industry employment. Rig counts provide an indication of both current employment levels related to well site preparation and drilling operations and future employment which would be required to **subsequently** operate producing wells.

Drilling and well completion processes for shale gas wells are significantly more labor intensive than those for conventional gas as they leverage complex horizontal drilling and multi-stage hydraulic fracturing processes. In fact, the Marcellus Shale Education and Training Center (MSETC) indicates that the direct workforce needed to drill a single well in the Marcellus Shale is comprised of over 410 individuals; translating to the equivalent of 11. 53 full-time direct jobs over the course of a year per well drilled. In contrast, postdrilling production operations for shale are similar to those used in conventional gas and require 0. 17 full-time **equivalent** employees per well. Based on these operational differences, and the actual proportion of industry rig counts focused on shale gas development in mid 2010, SBI Energy estimates about 24 percent (107,000) of overall oil and gas upstream employees were focused on shale gas development.

### Projected US Shale Gas Employment

In its 2010 Annual Energy Outlook, the EIA projects that US shale gas production will grow by 3. 2 percent per year between 2015 and 2020 reaching an annual output of 4. 51 Tcf, or 778 MMboe. In contrast, natural gas production from non-shale resources is projected to remain flat for the following five years. **Concurrently**, the rate of growth for crude oil production is also projected to be 1. 2 percent per year from 2015 to 2020.

Continued growth in drilling activity is required to enable this growth in US shale gas production and will support employment growth. SBI Energy projects upstream oil and gas industry employment related to shale gas will increase at an average annual growth rate of 5. 2 percent from 107,000 in 2010 to 138,000 in 2015. Employment levels are further projected to grow by 2. 4 percent annually to 156,000 in 2020.

A large number of jobs require various combinations of specific educational **credentials**, gas industry-specific training, certification, and field experience. However, shale gas development leverages thousands of employees in more traditional roles as well. These include all types of functions in areas such as administrative support, business, finance and accounting, information technology, and legal professionals.

### *Indirect andInduced Employment Resulting from US Shale Gas Development*

The economic and employment impacts of shale play development extend well beyond the direct employment of resources in upstream oil and gas. Upstream gas development activities **trigger** a chain of economic and employment impacts through further indirect and induced spending. Direct oil and gas **expenditures** on shale gas development are spent and respent in other sectors of the economy.

For example, activities and expenditures in gaining access to **mineral rights** create employment and economic opportunities before a well is even drilled. Mineral leasing generates demand for the services of geologists, surveyors, landmen, **attorneys**, title clerks, accountants, and a host of others to arrange and finalize contracts, which stimulates various industries outside of oil and gas. Further, over a million Americans own **royalty interests** in oil and gas through the private leasing of mineral rights and **royalty revenues** total billions of dollars annually. As individual wealth increases, this money **makes its way into** the economy in various ways and stimulates further employment in local economies.

In 2009, Penn State University examined the overall economic impact of shale gas development in the Pennsylvania Marcellus Shale for the prior year. The study found that while only about 2,100 direct jobs were created in the mining sector (of which the oil and gas industry is a part), over 27,000 additional jobs were created across a variety of industries as a result of the indirect and induced economic growth triggered by direct expenditures of shale gas developers. The biggest indirect and induced impacts were visible in the transportation and **warehousing** industry with the creation of almost 4,000 jobs. Further, the construction industry benefitted through the creation of nearly 3,800 jobs and health and social services employment increased by over 3,500. Shale gas dollars made their way into **manufacturing**, retail, finance and insurance, and real estate and supported the generation of over 800 jobs in each. Other studies focused on the impact of shale gas development in the Barnett, Haynesville, and Fayetteville plays have similarly found a significant **multiplier** effect on the level of overall employment generation.

## Words and Expressions

| | |
|---|---|
| trigger | 扳机;印发;触发 |
| recruit | 招聘;聘用;征募 |
| daunting | 使人畏缩的;使人气馁的 |
| rehabilitate | 使恢复;复兴 |
| compensation | 补偿;报酬;赔偿金 |
| lucrative | 有利可图的;赚钱的;合算的 |
| rig | 装备;钻探设备,钻机 |
| subsequently | 随后;后来 |
| equivalent | 等价物;相等的 |
| concurrently | 同时发生的 |
| credential | 证书;凭据 |
| induce | 诱导;引起;引诱 |
| expenditure | 支出;花费 |
| attorney | 律师;代理人 |
| warehousing | 仓储;把存入仓库 |
| manufacturing | 制造业;制造,生产 |
| multiplier | 乘数;增加者 |

## Phrases and Expressions

| | |
|---|---|
| Interstate Oil and Gas Compact Commission(IOGCC) | 州际石油天然气契约委员会 |
| well completion | 完井 |
| multi-stage hydraulic fracturing | 多级水力压裂 |
| mineral right | 矿产开发权 |
| royalty interest | 产区权益 |
| royalty revenues | 矿区收益 |
| make one's way into... | 进入 |

## Language Focus

1. Given the high levels of unemployment currently prevalent in the US, recruiting the number of employees needed to backfill retirees should not

be daunting.

(参考译文:鉴于美国目前失业率较高,招聘新职员来补充退休人员还没问题。)

本句中,given 引导的是过去分词短语作状语,given 有"考虑到"的意思。currently prevalent in the US 是形容词短语作后置定语。recruiting the number of employees 是动名词短语作句子的主语,should not be daunting 是谓语,needed to backfill retirees 是过去分词短语作后置定语。

2. Recent petroleum engineering graduates reportedly received starting salaries averaging about $75,000 in 2010, making petroleum engineering the most lucrative undergraduate degree this year.

(参考译文:据报道,2010 年石油工程毕业生起薪平均约达 75,000 美元,使得石油工程成为年度最赚钱的本科专业。)

本句中,graduates received... 为主谓结构,averaging about... 为现在分词短语作后置定语,修饰前面的 starting salaries; making... 为现在分词短语用作状语,表示结果,修饰整个句子。

3. As of July 2010, a total of 1,585 rotary rigs were actively drilling in the United States, 62 percent of which were drilling for gas.

(参考译文:2010 年 7 月,美国实际工作的旋转转机共有 1585 个,其中 62% 是用来开采天然气的。)

本句中 rigs were drilling 为主句主谓; 62 percent of which 引导定语从句,修饰 rotary rigs。为了句子结构的平衡,英语中定语从句常常与先行词分开。

## Reinforced Learning

### I. Answer the following questions for a comprehension of the text.

1. Describe the measures to attract talent and address the challenge of the widely publicized "big crew change."

2. What is the most lucrative undergraduate degree inAmerica according to the text?

3. Why do drilling and well completion processes for shale gas wells require more labors than conventional gas developing processes?

4. What're the requirements of the jobs in Shale Gas industry?

5. Describe the overall economic impact of shale gas development in the Pennsylvania Marcellus Shale for the prior year in short.

Ⅱ. **Multiple choice**：**choose the correct one from the alternative answers to give the exact meaning of the words.**

1. The employment gap caused by retiring workers is a <u>critical</u> strategic challenge for the industry，where the retirement age averages 55 years.

　A. faultfinding　　　B. crucial　　　C. analytical　　　D. correct

2. Industry organizations have successfully <u>lobbied</u> the federal government to increase internships for geosciences graduates and undergrads.

　A. persuaded　　　B. agreed　　　C. bribed　　　D. forced

3. All of these developments <u>indicate</u> a strong employment outlook for the industry over the next several years.

　A. demonstrate　　　B. dedicate　　　C. induct　　　D. indirect

4. Recent petroleum engineering graduates reportedly received starting salaries averaging about $75,000 in 2010，making petroleum engineering the most <u>lucrative</u> undergraduate degree this year.

　A. limitative　　　B. active　　　C. profitable　　　D. illuminative

5. Drilling and well completion processes for shale gas wells are significantly more labor <u>intensive</u> than those for conventional gas as they leverage complex horizontal drilling and multi-stage hydraulic fracturing processes.

　A. extensive　　　B. intense　　　C. ardent　　　D. concentrated

6. <u>Concurrently</u>，the rate of growth for crude oil production is also projected to be substantially lower than that of shale gas at 1.7 percent per year through 2015 and 1.2 percent per year from 2015 to 2020.

　A. simultaneously　　B. currently　　C. consciously　　D. continuously

7. A large number of jobs require various combinations of specific educational <u>credentials</u>，gas industry-specific training，certification，and field experience.

　A. recertification　　B. confidential　　C. certificates　　D. concretions

8. Direct oil and gas <u>expenditures</u> on shale gas development are spent and respent in other sectors of the economy.

　A. income　　　B. exclusives　　　C. exclusions　　　D. expenses

9. As individual wealth increases，this money makes its way into the economy in <u>various</u> ways and stimulates further employment in local economies.

　A. vicious　　　B. multifarious　　　C. furious　　　D. curious

10. Given the high levels of unemployment currently <u>prevalent</u> in the US，

recruiting the number of employees needed to backfill retirees should not be daunting.

    A. relevant        B. prevailing    C. popular        D. reverent

**Ⅲ. Multiple choice: read the four suggested translations and choose the best answer.**

1. Other studies focused on the impact of shale gas development in the Barnett, Haynesville, and Fayetteville plays have similarly found a significant <u>multiplier</u> effect on the level of overall employment generation.

    A. 复杂        B. 重要        C. 成双        D. 成倍

2. In the meantime, industry associations have conducted massive public relations campaigns designed to <u>rehabilitate</u> the industry's poor image.

    A. 居住        B. 建造        C. 修复        D. 回复

3. They provide incentives for workers approaching retirement age to stay with the company and leverage <u>compensation</u> incentives to attract new talent.

    A. 感谢        B. 报答        C. 薪酬        D. 补偿

4. In contrast, postdrilling production operations for shale are similar to those used in conventional gas and require 0. 17 full-time <u>equivalent</u> employees per well.

    A. 相似的        B. 相反的        C. 相对的        D. 相等的

5. However, shale gas development <u>leverages</u> thousands of employees in more traditional roles as well.

    A. 使用        B. 租用        C. 影响        D. 平衡

**Ⅳ. Put the following sentences into Chinese.**

1. However, the challenge lies in replacing the knowledge lost with the attrition of experienced workers in an industry where reaching a level of basic proficiency can require up to three years' field experience.

2. Recent petroleum engineering graduates reportedly received starting salaries averaging about $75,000 in 2010, making petroleum engineering the most lucrative undergraduate degree this year.

3. In contrast, post drilling production operations for shale are similar to those used in conventional gas and require 0. 17 full-time equivalent employees per well.

4. Continued growth in drilling activity is required to enable this growth in-

US shale gas production and will support employment growth.

5. As individual wealth increases, this money makes its way into the economy in various ways and stimulates further employment in local economies.

**V. Put the following paragraphs into Chinese.**

1. E&P and OFS companies have strengthened outreach programs to high schools and colleges and have developed mentoring, scholarship, and grant programs to support ongoing education for existing employees. They provide incentives for workers approaching retirement age to stay with the company and leverage compensation incentives to attract new talent.

2. The economic and employment impacts of shale play development extend well beyond the direct employment of resources in upstream oil and gas. Upstream gas development activities trigger a chain of economic and employment impacts through further indirect and induced spending. Direct oil and gas expenditures on shale gas development are spent and respent in other sectors of the economy.

# 2.3 Exploration and Production Industry Trends

## 🔲 Guidance to Reading

*Becauseof high prices for crude oil and natural gas, as well as the improvement of horizontal drilling and hydraulic fracturing technologies, E&P companies are able to realize profits from shale gas and seeking to further get access to substantial shale gas development. In North America, the landscape of shale gas production was usually dominated by independent E&P companies; however, as access to high-quality offshore and global conventional gas resources became increasingly limited and expensive, **IOCs** have begun to turn to shale gas resources to **replenish** reserves. In fact, approximately 75 percent of ExxonMobil's portfolio is accounted for by unconventional oil and gas resources. Hence, brilliant unconventional oil and gas exploration and production industry trends will be expected.*

## 🔲 Text

Over recent years, long-term opportunities presented by shale plays have attracted new participants to the shale gas industry. This activity has had the **dual** effects of **altering** the competitive landscape and **initiating** the transition of shale gas from a regionally-produced commodity to one which will potentially be produced globally. The successes in profitable shale gas recovery in North America have begun to attract the attention of international oil companies seeking to replenish reserves and foreign IOCs seeking to gain technical expertise in extracting from the unconventional resource.

Until very recently, shale gas exploration and production activities occurred mostly in theUnited States and, to a more limited degree, Canada. The landscape of shale producers was dominated by independent E&P companies. While **IOCs** focused instead on growth through offshore and non-domestic conventional gas resources and liquefied natural gas (**LNG**), these companies were able to realize profits from shale gas through the confluence of high prices for natural gas and the improvement of horizontal drilling and hydraulic fracturing technologies and further accumulate **substantial** shale asset **portfolios**.

As access to high-quality offshore and global conventional gas resources became increasingly limited and expensive, IOCs have begun to turn to shale gas resources to replenish reserves. Subsequently, prices for natural gas experienced a **prolonged** decrease, driven in large part by the success of the US shale plays, and several small to medium-sized operators began to seek strategic approaches to debt reduction.

**M&A** activities between independents and the energy majors in the US shale plays began an upward trend in 2008 and 2009. For example, BP USA bought into the Woodford and Fayetteville shales through **joint venture** partnerships with Chesapeake Energy followed by ExxonMobil's $41 billion acquisition of XTO Energy. This has been followed by continuing M&A activity between energy majors and independents, including a $2.25 billion joint venture which gave Total an ownership stake in the US shale assets of Chesapeake Energy, accelerating Total's expansion into gas, and the $4.7 billion cash purchase of East Resources by Royal Dutch Shell in May 2010.

Given this trend and the abundance of attractive shale gas producers already established in the market, it is likely that major energy companies will continue to purchase into shale gas resources in the United States; changing the landscape of shale producers substantially over the next several years.

**ExxonMobil**

*Corporate Overview*

Headquartered inIrving, Texas, ExxonMobil is one of the world's largest oil and gas companies. Their principal business is exploration and production of crude oil and natural gas, as well as the manufacture of petroleum products and transportation and sale of crude oil, natural gas, and petroleum products. ExxonMobil is the world's largest refiner and manufactures and markets commodity **petrochemicals**. Although the current company was formed on November 30, 1999, through the merger of Exxon and Mobil, its roots can be traced back through Standard Oil to 1870.

In June 2010, ExxonMobil concluded a $41 billion **acquisition** of XTO Energy, a US based E&P company specialized in unconventional natural gas. At the end of December 2009, XTO Energy reported proved reserves of 14. 8 trillion cubic feet of natural gas **equivalents** including shale gas, tight gas, coal-bed methane, and shale oil. Much of the holdings lie in three of the biggest shale gas formations in the US: the Marcellus, Haynesville, and Barnett shales. As a subsidiary of ExxonMobil, the company is named XTO Energy, Incorporated and it will focus on global development and production of unconventional resources. The acquisition represents a significant investment by ExxonMobil into shale gas assets and is expected to complement ExxonMobil's holdings in the United States, Canada, Germany, Poland, and Argentina.

*Product Portfolio*

ExxonMobil's proved hydrocarbon reserves were nearly 23 billion **barrels** of oil equivalent at the end of 2009. Nearly 50 percent of the company's proved reserves are in natural gas, with crude **condensates**, NGLs, **bitumen**, and **synthetic** oil accounting for the balance. Approximately 75 percent of the company's portfolio is accounted for by unconventional oil and gas resources. Additionally, the acquisition of **XTO Energy** is expected to further increase the company's participation in unconventional, shale gas production in the US.

ExxonMobil has been actively increasing its participation in unconventional resource development as a means of supplementing its **depleting** reserves of conventional natural gas over the past several years. The company has pursued tight gas plays in Colorado's Piceance Basin for many years. The company reports that it has a global unconventional resource base of approximately 70 trillion cubic feet. Additionally, the acquisition of XTO Energy adds shale-dominated unconventional gas resources of over 45 trillion cubic feet.

## BP

### *Corporate Background*

BP is involved in oil and natural gas exploration and production, processing and transportation, marketing and trading of natural gas, LNG and natural gas liquids (NGLs) as well as the refining, manufacturing, marketing, and transportation of crude oil and petrochemicals. BP's Alternative Energy business manages the company's activities in low-carbon energy, including solar, wind, biofuels, and carbon capture and **sequestration**. BP has operations in 80 countries around the world and is actively conducting exploration and production activities within 30 countries.

BP made a significant move into theUS shale gas market in 2008 when it acquired all of Chesapeake Energy's interests (90,000 acre leasehold) in the Woodford Shale play along with 135,000 acres in the Fayetteville Shale for $2.85 billion in cash. Further, the company has positions in the Haynesville and Eagle Ford plays. Together, the four plays represent a combined resource potential of over 10 Tcf. Internationally, **TNK-BP**, the company's joint venture in Russia, is seeking to begin shale gas exploration in Ukraine.

### *Product Portfolio*

BP's proved reserves, inclusive of BP subsidiaries and equity-accounted entities **exceed** 18 **MMboe**. Over 40 percent of the company's proved reserves are in natural gas; with crude, condensates, and NGLs comprising the balance. The company's total proved reserves have increased approximately 0.6 percent over the past five years, with crude oil reserves increasing 2.4 percent over the period while proved reserves of natural gas decreased 1.7 percent.

By the end of 2009, conventional sources accounted for over 62 percent of the company's total gas resources. Unconventional sources account for the balance and deliver about 10 percent of BP's natural gas production. BP notes that shale gas assets are becoming an increasingly important part of the company's gas business inNorth America. The company has access to more than 400 net acres in US shale plays including Haynesville, Fayetteville, Woodford, and Eagle Ford. Together, the four North American plays represent a combined resource potential of over 10 Tcf.

# Words and Expressions

| | |
|---|---|
| replenish | 补充;充满;添加(燃料等) |
| dual | 双重的;双倍的 |
| alter | 改变;更改;修改 |
| initiate | 开始;发起;使初步了解 |
| substantial | 大量的;很大程度的 |
| portfolio | 投资组合(商业);(产品或设计的)系列 |
| prolong | 延长;拖延 |
| petrochemicals | 石油化学产品 |
| acquisition | 收购;获得 |
| equivalent | 等价物,等值物;等量的,等值的 |
| barrel | 桶(石油计量单位,等于 42 加仑或 159 升) |
| condensate | 凝析油;冷凝物;浓缩物 |
| bitumen | 沥青;柏油 |
| synthetic | 合成的;人造的 |
| deplete | 用尽;消耗;使减少 |
| sequestration | 封存;隔离;扣押 |
| exceed | 超过(某数量);超越(限制、规定) |
| MMboe | 万桶油当量 |

# Phrases and Expressions

| | |
|---|---|
| International oil companies (IOCs) | 国际石油公司 |
| liquefied natural gas (LNG) | 液化天然气 |
| M&A | 企业并购;合并与收购 |
| joint venture | 合资企业 |
| XTO Energy | 克洛斯提柏能源公司 |
| TNK-BP | 秋明英国石油公司 |

# Language Focus

1. This activity has had the dual effects of altering the competitive landscape and initiating the transition of shale gas from a regionally-produced commodity to

one which will potentially be produced globally.

（参考译文：最近的页岩气开发产生了双重影响，一是改变了能源竞争格局，二是使页岩气从区域性产品开始转变为全球性产品。）

本句中 dual effects 分别指 altering the competitive landscape 和 initiating the transition of shale gas 两个动名词短语所表达的意思；transition...from...to... 为固定搭配。which 引导的是定语从句，修饰先行词 one，而 one 又指代前面的 commodity。

2. As access to high-quality offshore and global conventional gas resources became increasingly limited and expensive, IOCs have begun to turn to shale gas resources to replenish reserves.

（参考译文：由于全球常规天然气和海上高质量油气资源越来越有限，开发成本也越来越高，国际石油公司也开始转向页岩气开发，以增加油气储量。）

本句中 as 引导原因状语从句，附带说明"双方已知的原因"，语气比 since 弱，较为正式，位置较为灵活（常放于主句之前）。

2. Given this trend and the abundance of attractive shale gas producers already established in the market, it is likely that major energy companies will continue to purchase into shale gas resources in the United States.

（参考译文：鉴于这种趋势以及许多令人羡慕的页岩气生产商在市场上已初显成就，可以预计能源巨头将继续从美国购买页岩气资源。）

本句中 given 引导的短语作状语，意思是"鉴于；考虑到"。it is likely that... 为主谓结构，it 是形式主语，that 从句是真正主语。

3. Nearly 50 percent of the company's proved reserves are in natural gas, with crude condensates, NGLs, bitumen, and synthetic oil accounting for the balance.

（参考译文：将近百分之五十的已探明储量是天然气资源，其余部分为凝析油、天然气凝析液、沥青以及合成油。）

本句中 with 短语为独立主格结构作状语。balance 可译为"剩余部分"。

## 🔲 Reinforced Learning

### I. Answer the following questions for a comprehension of the text.

1. What have new players been interested in the shale gas industry?

2. When isshale gas production profitable?

3. Describe the proved oil and gas reserves of ExxonMobil and BP.

4. What's the principal business of ExxonMobil?

5. What does BP's Alternative Energy business involve?

**Ⅱ. Multiple choice: choose the correct one from the alternative answers to give the exact meaning of the words.**

1. This activity has had the dual effects of altering the competitive landscape and <u>initiating</u> the transition of shale gas from a regionally-produced commodity to one which will potentially be produced globally.

 A. irritate   B. lure   C. introduce   D. start

2. Until very recently, shale gas exploration and production activities occurred mostly in theUnited States and, to a more <u>limited</u> degree, Canada.

 A. local    B. imperfect C. finite    D. partial

3. While IOCs focused instead on growth through <u>offshore</u> and non-domestic conventional gas resources and liquefied natural gas (LNG).

 A. deep-sea  B. near-shore C. far-shore  D. land

4. As access to high-quality offshore and global conventional gas resources became increasingly limited and expensive, IOCs have begun to turn to shale gas resources to <u>replenish</u> reserves.

 A. supplement B. renew  C. replace   D. furnish

5. <u>Subsequently</u>, prices for natural gas experienced a prolonged decrease, driven in large part by the success of the US shale plays, and several small to medium-sized operators began to seek strategic approaches to debt reduction.

 A. successively B. afterwards C. therefore  D. next

6. Their <u>principal</u> business is exploration and production of crude oil and natural gas, as well as the manufacture of petroleum products and transportation and sale of crude oil, natural gas, and petroleum products.

 A. major   B. principle  C. minor   D. headmaster

7. <u>Approximately</u> 75 percent of the company's portfolio is accounted for by unconventional oil and gas resources.

 A. nearly   B. widely   C. minute   D. partly

8. Together, the four plays represent a combined resource <u>potential</u> of over 10 Tcf.

 A. prevalent  B. maybe   C. profound  D. possibility

9. The company has <u>pursued</u> tight gas plays in Colorado's Piceance Basin

for many years.

    A. pursuited      B. chased      C. trailed      D. engaged in

10. BP's proved reserves, inclusive of BP subsidiaries and equity accounted entities exceed 18 MMboe.

    A. excel      B. transcend      C. surpass      D. superior

**III. Multiple choice: read the four suggested translations and choose the best answer.**

1. These companies were able to realize profits from shale gas through the confluence of high prices for natural gas and the improvement of horizontal drilling and hydraulic fracturing technologies and further accumulate substantial shale asset portfolios.

    A. 意识到      B. 认清      C. 赚取      D. 成为现实

2. Nearly 50 percent of the company's proved reserves are in natural gas, with crude condensates, NGLs, bitumen, and synthetic oil accounting for the balance.

    A. 平衡      B. 抵消      C. 剩余部分      D. 差额

3. BP's Alternative Energy business manages the company's activities in low-carbon energy, including solar, wind, biofuels, and carbon capture and sequestration.

    A. 供选择的      B. 替代性的      C. 可再生的      D. 另外的

4. ExxonMobil's proved hydrocarbon reserves were nearly 23 billion barrels of oil equivalent at the end of 2009.

    A. 当量的      B. 等效的      C. 相似的      D. 模棱两可的

5. BP notes that shale gas assets are becoming an increasingly important part of the company's gas business in North America.

    A. 收入      B. 资产      C. 条件      D. 特点

**IV. Put the following sentences into Chinese.**

1. ExxonMobil has been actively increasing its participation in unconventional resource development as a means of supplementing its depleting reserves of conventional natural gas over the past several years.

2. Subsequently, prices for natural gas experienced a prolonged decrease, driven in large part by the success of the US shale plays, and several small to medium-sized operators began to seek strategic approaches to debt reduction.

3. As access to high-quality offshore and global conventional gas resources became increasingly limited and expensive, IOCs have begun to turn to shale gas resources to replenish reserves.

4. BP's Alternative Energy business manages the company's activities in low-carbon energy, including solar, wind, biofuels, and carbon capture and sequestration.

5. Nearly 50 percent of the company's proved reserves are in natural gas, with crude condensates, NGLs, bitumen, and synthetic oil accounting for the balance.

**V. Put the following paragraphs into Chinese.**

1. Until very recently, shale gas exploration and production activities occurred mostly in theUnited States and, to a more limited degree, Canada. The landscape of shale producers was dominated by independent E&P companies. While IOCs focused instead on growth through offshore and non-domestic conventional gas resources and liquefied natural gas (LNG), these companies were able to realize profits from shale gas through the confluence of high prices for natural gas and the improvement of horizontal drilling and hydraulic fracturing technologies and further accumulate substantial shale asset portfolios.

2. Given this trend and the abundance of attractive shale gas producers already established in the market, it is likely that major energy companies will continue to purchase into shale gas resources in the United States; changing the landscape of shale producers substantially over the next several years.

# 2.4 Oilfield Services Industry Trends

## Guidance to Reading

*Oilfield services companies support the upstream petroleum sector and derive revenues from providing technology, consulting, project management, well services and maintenance, and information solutions to the oil and gas exploration and production industry. Unconventional oil and gas resources have a great potential, playing a vital role in the global energy landscape, and making the key sector in the future of petroleum industry development. Consequently, the brilliant unconventional oil and gas exploration and production industry trends will also bring about attractive OFS industry trends. The two largest OFS com-*

*panies in the world Schlumberger and Halliburton will be introduced in this section*

## 卐 Text

**Oilfield services**（OFS）companies provide technology, consulting, project management, well services and maintenance, and information solutions to the oil and gas exploration and production industry. The customers of OFS providers include independent exploration and production companies and national and international oil companies, which depend on oilfield services firms for the equipment, **infrastructure**, and technology to extract, store, and transport oil and natural gas.

The OFS industry is **fragmented**, but can be generally characterized into four types of participants: large, **integrated**, and global players like Schlumberger and Halliburton; small, specialized providers of **niche** technology, software, engineering services, or specialized equipment; low-tech or commodity providers of low-cost equipment; and regional participants. The sector has experienced a trend in recent years toward the development of diverse and comprehensive product and services portfolios, increasing participants' abilities to offer **bundled** product, service, and integration options, resulting in fewer complications and lower cost for clients. In line with this trend, the OFS sector continued to experience recent M&A activity targeted at rounding out providers' unconventional gas service capabilities.

For example, in April 2010, Baker Hughes closed an acquisition of BJ Services for $5.5 billion; **citing** the company's pressure-pumping equipment and other services used in shale gas and other unconventional gas formations as key reasons for the purchase. Further, the shale gas expertise of former joint-venture partner Smith International was a key factor in Schlumberger's August 2010 $11 billion all-paper acquisition of the company. Smith International is the world's second-largest provider of **drill bits**; an equipment market to which Schlumberger previously had no access. The acquisition adds Smith's *I-Drill*, a computer modeling system that analyzes drilling operations and predicts drilling performance to the Schlumberger portfolio. Halliburton has also been acquiring companies with technologies to better support unconventional oil and gas assets over the past 24 months including Pinnacle Technologies, Evans Engineering, Permedia Research Group, and Boots & Coots, among others.

### Schlumberger, Ltd

*Corporate Overview*

Schlumberger was founded in 1926 and is the world's largest oilfield services company. In 2009, the company maintained two business **segments**: Oilfield Services and Western Geco. The Oilfield Services segment provides products and services in support of oil and gas development projects. Western Geco is a surface **seismic** acquisition and processing company.

In August 2010, the company acquired Smith International, Inc. the world's second-largest provider of drill bits. The deal simultaneously provided Schlumberger access to highly-rated drillbits, the ability to close a gap in its product/service portfolio, and full ownership in M-I SWACO, formerly a joint venture with Smith. M-I SWACO provides drilling and completion fluids to **facilitate** minimal formation damage during completion and workover operations.

The company operates 22 research and engineering centers for the development of advanced technology programs. Technology development focuses on enhanced oilfield efficiency, reducing finding and production costs, productivity improvement, maximization of reserve recovery, and asset value improvement. Schlumberger invented **wireline logging**, a technique for obtaining downhole data in oil and gas wells. Further, it is an industry leader in **well testing**, measurement-while-drilling, **logging-while-drilling**, directional drilling services, fully computerized logging, and **geoscience** software. The company employs about 105,000 and operates in 80 countries; with principal offices located in Paris, Houston, and The Hague.

### *Product Portfolio*

Schlumberger supports oil and gas development projects from exploration to production through various products offered through its Oilfield Services segment. These include wireline products and services, drilling and measurements, completions, and well services. Software, information management, and IT infrastructure services to support various oil and gas operational processes are additionally offered through the business segment.

TheWestern Geco business segment provides comprehensive reservoir imaging, monitoring, and development services. Services range from 3D and time-lapse (4D) seismic surveys to multi-component surveys for **delineating** prospects and reservoir management. Further, the Western Geco business segment maintains extensive seismic data processing centers to support complex data processing projects.

## Halliburton Company

### *Corporate Overview*

Halliburton, the second-largest oilfield services company globally, was established in 1919. The company provides upstream oil and gas companies with a complete range of services, from the location of hydrocarbons to production optimization through the life of the well. Halliburton's broad portfolio, and the ability to offer integrated-package services, has enabled the company to capture a substantial portion of the market for oilfield services as customers seek to manage total project costs by contracting large packages of services from one vendor.

The company **is credited with** the commercialization of hydraulic fracturing in the 1940s and **was instrumental in** pioneering the horizontal drilling and stimulation methodologies used in the Barnett Shale beginning in the 1980s. Halliburton has provided various services in all active major shale basins since. It is a leading technology provider to oil and gas companies working in unconventional plays and has been highly **acquisitive** in recent years. Several of the company's recent acquisitions include:

- Pinnacle Technologies, provider of real time tiltmeter and microseismic mapping and reservoir monitoring services,

- Evans Engineering, Inc. , with wireline and perforating products,

- Permedia Research Group, suppliers of petroleum systems modeling software and services, and

- Boots & Coots, Inc. , well intervention services and pressure control products.

### *Product Portfolio*

Halliburton maintains two business segments: Drilling and Evaluation and Completion and Production. Equipment and services to evaluate reservoirs and model, measure, and optimize well placement and stability are offered from the Drilling and Evaluation segment. Products and services include **reservoir modeling**, drilling and fluid services, drill bits, wireline and **perforating** services, testing and **subsea** services, software and asset solutions, and project management services. **Well cementing**, completion, **stimulation**, and intervention tools and services are offered by the Completion and Production business unit.

## Words and Expressions

| | |
|---|---|
| infrastructure | 基础设施;公共建设 |
| fragment | 分裂;破碎;使成碎片 |
| integrated | 综合的;整体的;完整的 |
| niche | 利基;有利可图的市场(或形势等) |
| bundle | 捆绑;包;束 |
| cite | 引证;引用 |
| segment | 部分;段 |
| seismic | 地震的;地震引起的 |
| facilitate | 促进;帮助;使容易 |
| geoscience | 地球科学 |
| delineate | 描绘;描述;划定 |
| acquisitive | 想获得的,贪得的 |
| tiltmeter | 测斜仪 |
| perforate | 射孔;穿孔 |
| subsea | 海底的;在海底 |
| stimulation | 增产 |

## Phrases and Expressions

| | |
|---|---|
| Oilfield services (OFS) | 油田服务 |
| drill bit | 钻头 |
| wireline logging | 电缆测井 |
| well testing | 试井 |
| logging-while-drilling | 随钻测井 |
| be credited with | 归功于,认为 |
| be instrumental in | 有助于;在某方面起作用 |
| reservoir modeling | 油藏模拟 |
| well cementing | 固井 |

## Language Focus

1. Customers of OFS providers include independent exploration and produc-

tion companies and national and international oil companies, which depend on oil-field services firms for the equipment, infrastructure, and technology to extract, store, and transport oil and natural gas.

（参考译文：油田服务公司的客户包括独立勘探开发公司、国家石油公司和国际石油公司，这些公司需要油服公司提供基础设施、设备和技术来进行油气开采、储存和运输。）

本句中，which 引导的是非限制性定语从句，修饰前面句中提到的系列公司，"depend on"是固定搭配，可译为"取决于，依靠"。depend on sb. for...可译为"依靠某人得到……；指望某人得到……"。

2. The sector has experienced a trend in recent years toward the development of diverse and comprehensive product and services portfolios, increasing partici-pants' abilities to offer bundled product, service, and integration options, resulting in fewer complications and lower cost for clients.

（参考译文：最近几年，该行业向着产品和服务多元化综合趋势发展，加强集产品和服务于一身的一体化服务能力，令客户享受更便捷的低成本服务。）

本句中 increasing 和 resulting 引导的是两个现在分词短语，在句中作状语，表示结果。

3. The acquisition adds Smith's I-Drill, a computer modeling system that an-alyzes drilling operations and predicts drilling performance to the Schlumberger portfolio.

（参考译文：这次收购使史密斯国际的计算机建模系统 I-Drill 也进入斯伦贝谢的资产组合当中，这一系统的功能是对钻井作业进行分析和预测。）

本句中 a computer modeling system 是 Smith's I-Drill 的同位语，system 后面有一个以 that 引导的定语从句，该定语从句中有两个并列的谓语 analyzes 和 predicts。

4. The company employs about 105,000 and operates in 80 countries; with principal offices located in Paris, Houston, and The Hague.

（参考译文：该公司业务遍布80个国家，在职员工有10.5万名；在巴黎、休斯顿和海牙都设有总部。）

本句中包含"with + n. /pron. + done"的结构，此时，过去分词和前面的名词是逻辑上的主谓关系。这是一种由 with 引导的独立主格结构。

5. It is a leading technology provider to oil and gas companies working in un-conventional plays and has been highly acquisitive in recent years.

（参考译文:哈里伯顿是为非常规油气开发公司提供技术的主要供应商,最近几年发展势头大增。）

本句中,working in... plays 是现在分词短语,在句中作定语,修饰前面的 oil and gas companies。

## Reinforced Learning

**I. Answer the following questions for a comprehension of the text.**

1. What services doOFS companies provide?

2. Which company is the world's second-largest provider of drill bits?

3. Which company is the world's largest oilfield services company, and how many business segments does it maintain?

4. What does Schlumberger's technology development focus on?

5. What are the two business segments which Halliburton maintains?

**II. Multiple choice:choose the correct one from the alternative answers to give the exact meaning of the words.**

1. Oilfield services ( OFS ) companies provide technology, consulting, project management, well services and maintenance, and information solutions to the oil and gas exploration and production industry.

 A. constraining  B. consuming  C. contending  D. conferring

2. The OFS industry is fragmented, but can be generally characterized into four types of participants.

 A. contestants      B. partners

 C. members       D. accomplices

3. The sector has experienced a trend in recent years toward the development of diverse and comprehensive product and services portfolios.

 A. overall    B. compulsive  C. inclusive   D. substantial

4. In line with this trend, the OFS sector continued to experience recent M&A activity targeted at rounding out providers' unconventional gas service capabilities.

 A. in accordance with    B. owing to

 C. in terms of      D. according to

5. For example, in April 2010, Baker Hughes closed an acquisition of BJ Services for $5.5 billion; citing the company's pressure-pumping equipment

and other services used in shale gas and other unconventional gas formations as key reasons for the purchase.

A. conceal     B. mention     C. introduce     D. refer to

6. Western Geco is a surface seismic acquisition and processing company.

A. property     B. obtainment     C. acquaintance     D. knowledge

7. Software, information management, and IT infrastructure services to support various oil and gas operational processes are additionally offered through the business segment.

A. supplementarily     B. therefore     C. then     D. extraly

8. The company is credited with the commercialization of hydraulic fracturing in the 1940s and was instrumental in pioneering the horizontal drilling and stimulation methodologies used in the Barnett Shale beginning in the 1980s.

A. has devoted to     B. is thought to have realized

C. is regarded as     D. has done contribution to

9. Equipment and services to evaluate reservoirs and model, measure, and optimize well placement and stability are offered from the Drilling and Evaluation segment.

A. estimate     B. equate     C. evoke     D. assess

10. Well cementing, completion, stimulation, and intervention tools and services are offered by the Completion and Production business unit.

A. interval     B. interruption     C. interference     D. workover

**III. Multiple choice: read the four suggested translations and choose the best answer.**

1. The OFS industry is fragmented, but can be generally characterized into four types of participants.

A. 以……为表率     B. 性格

C. 字母     D. 特征

2. In 2009, the company maintained two business segments: Oilfield Services and Western Geco.

A. 维修;保养     B. 坚持认为

C. 与……保持联系     D. 维持;经营

3. The deal simultaneously provided Schlumberger access to highly-rated drillbits, the ability to close a gap in its product/service portfolio, and full ownership in M-I SWACO, formerly a joint venture with Smith.

A. 同时地,同步地 B. 同时代地     C. 自发地      D. 同一地

4. The company is credited with the <u>commercialization</u> of hydraulic fracturing in the 1940s and was instrumental in pioneering the horizontal drilling and stimulation methodologies used in the Barnett Shale beginning in the 1980s.

A. 公司化      B. 商业化      C. 服务化      D. 贸易化

5. Well cementing, completion, <u>stimulation</u>, and intervention tools and services are offered by the Completion and Production business unit.

A. 兴奋      B. 刺激;激励     C. 增产      D. 积累

## IV. Put the following sentences into Chinese.

1. Oilfield services (OFS) companies provide technology, consulting, project management, well services and maintenance, and information solutions to the oil and gas exploration and production industry.

2. In line with this trend, the OFS sector continued to experience recent M&A activity targeted at rounding out providers' unconventional gas service capabilities.

3. The acquisition adds Smith's I-Drill, a computer modeling system that analyzes drilling operations and predicts drilling performance to the Schlumberger portfolio.

4. The Oilfield Services segment provides products and services in support of oil and gas development projects. Western Geco is a surface seismic acquisition and processing company.

5. The company provides upstream oil and gas companies with a complete range of services, from the location of hydrocarbons to production optimization through the life of the well.

## V. Put the following paragraphs into Chinese.

1. The company operates 22 research and engineering centers for the development of advanced technology programs. Technology development focuses on enhanced oilfield efficiency, reducing finding and production costs, productivity improvement, maximization of reserve recovery, and asset value improvement. Schlumberger inventedwireline logging, a technique for obtaining downhole data in oil and gas wells. Further, it is an industry leader in well testing, measurement-while-drilling, logging-while-drilling, directional drilling services, fully computerized logging, and geoscience software.

2. Halliburton's broad portfolio, and the ability to offer integrated-package services, has enabled the company to capture a substantial portion of the market for oilfield services as customers seek to manage total project costs by contracting large packages of services from one vendor. The company is credited with the commercialization of hydraulic fracturing in the 1940s andwas instrumental in pioneering the horizontal drilling and stimulation methodologies used in the Barnett Shale beginning in the 1980s. Halliburton has provided various services in all active major shale basins since.

# Chapter 3  Unconventional Gas Development

## 3.1  Shale and Tight Gas Development (1)

### 🔲 Guidance to Reading

*Shale gas and tight gas are natural gas produced from reservoir rocks with such low permeability that massive well drilling and hydraulic fracturing are necessary to produce the well at economic rates, which will be certainly destructive to the environment. In North America, there has recently been a move towards drilling multiple wells from a single site, or pad, in order to limit the amount of disruption of well construction. As the rate of gas flow determines directly the cash flow from the well, and low flow rates cannot justify the operating expenses and provide a return on the capital invested, more measures to increase the flow of hydrocarbons to the well are necessary.*

### 🔲 Text

Shales are geological rock formations rich in clays, typically derived from fine sediments, deposited in fairly quiet environments at the bottom of seas or lakes, having then been buried over the course of millions of years. When a significant amount of organic matter has been deposited with the sediments, the shale rock can contain organic solid material called **kerogen**. If the rock has been heated up to sufficient temperatures during its burial history, part of the kerogen will have been transformed into oil or gas (or a mixture of both), depending on the temperature conditions in the rock.

This transformation typically increases pressure within the rock, resulting in part of the oil and gas being **expelled** from the shale and migrating upwards into other rock formations, where it forms conventional oil and gas reservoirs. The shales are the source rock for the oil and gas found in such conventional reservoirs. Some, or occasionally all, of the oil and gas formed in the shale can remain trapped there, thus forming shale gas or light tight oil reservoirs.

Tight gas reservoirs originate in the same way as conventional gas reservoirs:

the rock into which the gas migrates after being expelled from the source rock just happens to be of very low **permeability**. As a result, tight gas reservoirs also require special techniques to achieve commercial flow rates. On the other hand, they tend to have better recovery factors than shale gas deposits and, therefore, higher density of **recoverable hydrocarbons** per unit of surface area.

### Well construction

The drilling phase is the most visible and **disruptive** in any oil and gas development-particularly so in the case of shale gas or tight gas because of the larger number of wells required. On land, a drilling rig, associated equipment and pits to store drilling fluids and waste typically occupy an area of 100 metres by 100 metres (the well site). Setting up drilling in a new location might involve between 100 and 200 truck movements to deliver all the equipment, while further truck movements will be required to deliver supplies during drilling and completion of the well.

Each well site needs to be chosen **taking account** not only **of** the subsurface geology, but also of a range of other concerns, including **proximity** to populated areas and existing infrastructure, the local ecology, water availability and **disposal** options, and seasonal restrictions related to climate or wildlife concerns. In North America, there has recently been a move towards drilling **multiple wells** from a single site, in order to limit the amount of disruption and thereby the overall environmental impact of well construction. In 2011, according to industry sources, around 30% of all new shale and tight gas wells in the United States and Canada were multiple wells drilled from pads.

Once drilling starts, it is generally a 24-hour-per-day operation, creating noise and **fumes** from diesel generators, requiring lights at night and creating a regular stream of truck movements during **mobilisation/demobilisation** periods. Drilling operations can take anything from just a few days to several months, depending on the depth of the well and type of rock encountered. As the drill bit bores through the rock, **drilling fluid** known as "**mud**" is circulated through the wellbore in order, among other tasks, to control pressure in the well and remove cuttings created by the drill bit from the well. This **lubricating** "mud" consists of a base fluid, such as water or oil, mixed with salts and solid particles to increase its density and a variety of chemical additives. Mud is stored either in mobile containers or in open pits which are dug into the ground and **lined with** impermeable material. The volume of material in the pits needs to be monitored and contained to prevent leaks or spills. A drilling rig might have several hundred tonnes of mud in use at any one time, which creates a large demand for supplies. Once used, the mud must be either recycled or disposed of safely. Rock cuttings recovered from the mud during the drilling process amount to between 100 and 500 tonnes per well, depending on the depth. These, too, need to be disposed of in an environmentally acceptable fashion.

A combination of steel casing and cement in the well provides an essential barrier to ensure that high-pressure gas or liquids from deeper down cannot escape into shallower rock formations or water **aquifers**. This barrier has to be designed to withstand the cycles of stress it will endure during the subsequent hydraulic fracturing, without suffering any cracks. The design aspects that are most important to ensure a leak-free well include the drilling of the well bore to specification (without additional twists, turns or **cavities**), the positioning of the casing in the centre of the well bore before it is cemented in place (this is done with **centralisers** placed at regular intervals along the casing as it is run in the hole, to keep it away from the rock face) and the correct choice of cement. The cement design needs to be studied both for its liquid properties during pumping (to ensure that it gets to the right place) and then for its mechanical strength and flexibility, so that it remains intact. The setting time of the cement is also a critical factor-cement that takes too long to set may have reduced strength; equally, cement that sets before it has been fully pumped into place requires difficult remedial action.

### Well completion

Once the well has been drilled, the final casing cemented in place across the **gas-bearing rock** has to be perforated in order to establish communication between the rock and the well. The pressure in the well is then lowered so that hydrocarbons can flow from the rock to the well, driven by the pressure differential. With shale and tight gas, the flow will be very low, because of the low permeability of the rock. As the rate of hydrocarbon flow determines directly the cash flow from the well, low flow rates can mean there is insufficient revenue to pay for operating expenses and provide a return on the capital invested. Without additional measures to accelerate the flow of hydrocarbons to the well, the operation is then not economic.

## 🔲 Words and Expressions

| | |
|---|---|
| shales | |
| kerogen | 干酪根 |
| expel | 排出;逐出 |
| permeability | 渗透率;渗透性 |
| disruptive | 破坏性的;分裂性的;制造混乱的 |
| proximity | 亲近;接近;临近 |
| disposal | 处置;处理;清理 |
| fume | 烟雾;烟气;废气 |
| mobilisation | 调动;转移 |
| demobilisation | 遣散;出场 |
| lubricating | 滑的,涂油;加润滑油的 |
| aquifer | 蓄水层;含水层 |
| cavity | 腔;洞,凹处 |
| centralizer | 扶正器 |

## 🔲 Phrases and Expressions

| | |
|---|---|
| recoverable hydrocarbons | 可采油气 |
| drilling fluid/mud | 钻井液 |
| taking account of... | 考虑 |
| be lined with... | 衬有... |
| multiple wells | 多底井 |
| gas-bearing rock | 含气岩层 |

## 🔲 Language Focus

1. Shales are geological rock formations rich in clays, typically derived from fine sediments, deposited in fairly quiet environments at the bottom of seas or lakes, having then been buried over the course of millions of years.

（参考译文:页岩是富含黏土的地质岩层,通常由埋藏于海底或湖底几百万年的细粒沉积物形成。）

本句是简单句。主谓结构是 Shales are geological rock formations,表语 geological rock formations 的后面有两个定语,一个是形容词短语 rich in clays,另一个是过去分词短语 derived from...;sediments 后又带有两个自己的后置定语 deposited in...,和 having been...。

2. If the rock has been heated up to sufficient temperatures during its burial history,part of the kerogen will have been transformed into oil or gas ( or a mixture of both),depending on the temperature conditions in the rock.

[参考译文:如果沉积过程中地层温度足够高,部分干酪根便会转化为石油或天然气(或两者的混合物)。]

本句是含有 if 引导的条件状语从句的复合句,主句用将来完成时,意为"假如在那种条件下,就会产生了……结果"。depending on the temperature conditions in the rock 是现在分词短语做伴随状语。

3. This transformation typically increases pressure within the rock,resulting in part of the oil and gas being expelled from the shale and migrating upwards into other rock formations,where it forms conventional oil and gas reservoirs.

(参考译文:这一转化过程通常会使页岩储层压力增高,部分石油和天然气将向上运移至其他岩层,最终形成常规油气藏,因而页岩是常规油气藏的生油岩。)

本句是含有定语从句的复合句,where 引导的定语从句修饰 other rock formations。主句中的现在分词短语 resulting in part of the oil and gas being expelled from the shale and migrating upwards into other rock formations 作结果状语,在这个分词短语中,part of the oil and gas being expelled from the shale and migrating upwards into other rock formations 是独立主格结构,part of the oil and gas 和 being... 及 migrating... 是逻辑上的主谓关系,作 resulting in 的宾语。

4. Setting up drilling in a new location might involve between 100 and 200 truck movements to deliver all the equipment,while further truck movements will be required to deliver supplies during drilling and completion of the well.

(参考译文:到另一地点进行钻井作业时,单钻井设备就需要卡车来回运送 100~200 趟,而且钻井和完井过程中还需卡车运送物资。)

本句为复合句,动名词短语 setting up drilling in a new location 是主句主语。While 引导从句表示对比。

5. Without additional measures to accelerate the flow of hydrocarbons to the well,the operation is then not economic.

(参考译文为:如果不采取进一步的增产措施,开采作业就无法带来经济效益。)

本句含有"without + 名词短语 + 不定式"的独立主格结构,做句子的状语。

## 🔲 Reinforced Learning

### I. Answer the following questions for a comprehension of the text.

1. What are shale formations like?

2. Why is the drilling phase disruptive in shale gas development?

3. What does the "mud" consist of?

4. What's the purpose of the combination of steel casing and cement in the well?

5. What's the next procedure after the well has been drilled?

### II. Multiple choice: choose the correct one from the alternative answers to give the exact meaning of the words.

1. Shales are geological rock formations rich in clays, typically <u>derived</u> from fine sediments.

A. obtained     B. originated     C. designated     D. detached

2. When a <u>significant</u> amount of organic matter has been deposited with the sediments, the shale rock can contain organic solid material called kerogen.

A. considerable    B. important     C. profound     D. available

3. If the rock has been heated up to <u>sufficient</u> temperatures during its burial history, part of the kerogen will have been transformed into oil or gas (or a mixture of both), depending on the temperature conditions in the rock.

A. competent     B. enough     C. incomplete     D. successive

4. This transformation typically increases pressure within the rock, resulting in part of the oil and gas being <u>expelled</u> from the shale and migrating upwards into other rock formations, where it forms conventional oil and gas reservoirs.

A. banished     B. exiled     C. dismissed     D. removed

5. On the other hand, they tend to have better recovery factors than shale gas <u>deposits</u> and, therefore, higher density of recoverable hydrocarbons per unit of surface area.

A. accumulation    B. overdrafts     C. loans     D. repayments

6. The drilling phase is the most visible and <u>disruptive</u> in any oil and gas development-particularly so in the case of shale gas or tight gas because of the

larger number of wells required.

    A. unsetting     B. riotous     C. destructive     D. factious

7. Each well site needs to be chosen taking account not only of the subsurface geology, but also of a range of other concerns, including proximity to populated areas and existing infrastructure, the local ecology, water availability and disposal options, and seasonal restrictions related to climate or wildlife concerns.

    A. administration   B. clearance     C. elimination     D. treating

8. This barrier has to be designed to withstand the cycles of stress it will endure during the subsequent hydraulic fracturing, without suffering any cracks.

    A. endure     B. defy     C. repel     D. abide

9. The cement design needs to be studied both for its liquid properties during pumping (to ensure that it gets to the right place) and then for its mechanical strength and flexibility, so that it remains intact.

    A. in good shape   B. incomplete     C. integrate     D. marred

10. As the rate of hydrocarbon flow determines directly the cash flow from the well, low flow rates can mean there is insufficient revenue to pay for operating expenses and provide a return on the capital invested.

    A. compensation   B. yield     C. expenditure     D. income

**III. Multiple choice: read the four suggested translations and choose the best answer.**

1. These, too, need to be disposed of in an environmentally acceptable fashion.

    A. 时尚     B. 时装     C. 方式     D. 改变

2. Tight gas reservoirs originate in the same way as conventional gas reservoirs: the rock into which the gas migrates after being expelled from the source rock just happens to be of very low permeability.

    A. 运移     B. 移植     C. 迁徙     D. 移居

3. The drilling phase is the most visible and disruptive in any oil and gas development-particularly so in the case of shale gas or tight gas because of the larger number of wells required.

    A. 位相     B. 行为     C. 阶段     D. 逐步执行

4. Each well site needs to be chosen taking account not only of the subsurface geology, but also of a range of other concerns, including proximity to popu-

lated areas and existing infrastructure, the local ecology, water availability and disposal options, and seasonal restrictions related to climate or wildlife concerns.

    A. 接近        B. 远离        C. 疏远        D. 相似

5. This lubricating "mud" consists of a base fluid, such as water or oil, mixed with salts and solid particles to increase its density and a variety of chemical additives.

    A. 添加剂        B. 防腐剂        C. 调味剂        D. 增白剂

**Ⅳ. Put the following sentences into Chinese.**

1. This transformation typically increases pressure within the rock, resulting in part of the oil and gas being expelled from the shale and migrating upwards into other rock formations, where it forms conventional oil and gas reservoirs.

2. A combination of steel casing and cement in the well provides an essential barrier to ensure that high-pressure gas or liquids from deeper down cannot escape into shallower rock formations or water aquifers.

3. The drilling phase is the most visible and disruptive in any oil and gas development-particularly so in the case of shale gas or tight gas because of the larger number of wells required.

4. Once drilling starts, it is generally a 24-hour-per-day operation, creating noise and fumes from diesel generators, requiring lights at night and creating a regular stream of truck movements during mobilization/demobilization periods.

5. The cement design needs to be studied both for its liquid properties during pumping (to ensure that it gets to the right place) and then for its mechanical strength and flexibility, so that it remains intact.

**Ⅴ. Put the following paragraphs into Chinese.**

1. When a significant amount of organic matter has been deposited with the sediments, the shale rock can contain organic solid material called kerogen. If the rock has been heated up to sufficient temperatures during its burial history, part of the kerogen will have been transformed into oil or gas (or a mixture of both), depending on the temperature conditions in the rock.

2. With shale and tight gas, the flow will be very low, because of the low permeability of the rock. As the rate of hydrocarbon flow determines directly the cash flow from the well, low flow rates can mean there is insufficient revenue to

pay for operating expenses and provide a return on the capital invested. Without additional measures to accelerate the flow of hydrocarbons to the well, the operation is then not economic.

## 3.2　Shale and Tight Gas Development (2)

### 🔲 Guidance to Reading

*Hydraulic fracturing and acidizing are two essential measures to increase the flow of hydrocarbons in shale and tight gas production. The technique where only acid is injected is referred to as acidizing. The first experimental use of hydraulic fracturing was in 1947, and the first commercially successful applications were in 1949. Hydraulic fracturing as a well stimulation is usually conducted once in the life of the well and greatly enhances fluid removal and well productivity, but there has been an increasing trend towards multiple hydraulic fracturing as production declines. However, the resulting environmental risks cannot be ignored, including contamination of ground water, depletion of fresh water, contamination of the air, noise pollution, and surface contamination from spills and flow-back.*

### 🔲 Text

#### *Acid treatment and hydraulic fracturing*

Several technologies have been developed over the years to enhance the flow from lowpermeability reservoirs. Acid treatment, involving the **injection** of small amounts of strong acids into the reservoir to dissolve some of the rock minerals and enhance the permeability of the rock near the **wellbore**, is probably the oldest and is still widely practised, particularly in **carbonate** reservoirs. Wells with long horizontal or lateral sections (known as horizontal wells) can increase dramatically the contact area between the reservoir rock and the wellbore, and are likewise effective in improving project economics. Hydraulic fracturing, developed initially in the late 1940s, is another effective and commonly-practised technology for low-permeability reservoirs. When rock permeability is extremely low, as in the case of shale gas or light tight oil, it often takes the combination of horizontal wells and hydraulic fracturing to achieve commercial rates of production. Advances in the application of these two techniques, in combination, largely explain the surge in shale gas production in the United States since 2005.

Hydraulic fracturing involves pumping a fluid-known as fracturing fluid-at high pressure into the well and then, far below the surface, into the surrounding target rock. This creates fractures or **fissures**a few millimeters wide in the rock. These fissures can extend tens or, in some cases, even hundreds of meters away from the well bore. Once the pressure is released, these fractures would tend to close again and not produce any lasting improvement in the flow of hydrocarbons. To keep the fractures open, small particles, such as sand or **ceramic beads**, are added to the pumped fluid to fill the fractures and to act as **proppants**, i. e. they **prop open** the fractures thus allowing the gas to escape into the well.

In many cases, a series of fractures is created at set intervals, one after the other, about every 100 meters along the horizontal wellbore. This **multi-stage fracturing** technique has played a key role in unlocking production of shale gas and light tight oil in the United States and promises to do likewise elsewhere in the world. A standard **single-stage** hydraulic **fracturing** may pump down several hundred cubic meters of water together with prop pant and a mixture of various **chemical additives**. In shale gas wells, a multi-stage fracturing would commonly involve between ten and twenty stages, multiplying the volumes of water and solids by 10 or 20, and hence the total volumes for water use might reach from a few thousand to up to twenty thousand cubic metres of water per well and volumes of prop pant of the order of 1 000 to 4 000 tones per well. The repeated stresses on the well from **multiple** high-pressure procedures increase the **premium** on good well design and construction to ensure that gas bearing formations are completely isolated from other strata penetrated by the well.

Once the hydraulic fracturing has been completed, some of the fluid injected during the process flows back up the well as part of the produced stream, though typically not all of it-some remains trapped in the treated rock. During this flow-back period, typically over days (for a single-stage fracturing) to weeks (for a multi-stage fracturing), the amount of flow back of fracturing fluid decreases, while the hydrocarbon content of the produced stream increases, until the flow from the well is primarily hydrocarbons.

Best practice during this period is to use a so-called "**green completion**" or "**reduced emissions completion**", whereby the hydrocarbons are separated from the fracturing fluid (and then sold) and the residual flow-back fluid is collected for processing and recycling or disposal. However, while collecting and processing the fluid is standard practice, capturing and selling the gas during this initial flow-back phase requires investment in gas separation and processing facilities, which does not always take place. In these cases, there can be **venting** of gas to the atmosphere (mostly methane, with a small fraction of VOCs) or flaring (burning) of hydrocarbon or hydrocarbon/water mixtures. Venting and/or **flaring** of the gas at this stage are the main reasons why shale and tight gas can give rise to higher greenhouse-gas emissions than conventional production.

### Production

Once wells are connected to processing facilities, the main production phase can begin. During production, wells will produce hydrocarbons and waste streams, which have to be managed. But the well site itself is now less visible: a "**Christmas tree**" of valves, typically one metre high, is left on top of the well, with production being piped to processing facilities that usually serve several wells. In some cases, the operator may decide to repeat the hydraulic fracturing procedure at later times in the life of the producing well, a procedure called re-fracturing. This was more frequent in vertical wells but is currently relatively rare in horizontal wells, occurring in less than 10% of the horizontal shale-gas wells drilled in the United States.

The production phase is the longest phase of the lifecycle. For a conventional well, production might last 30 years or more. For an unconventional development, the productive life of a well is expected to be similar, but shale gas wells typically exhibit a burst of initial production and then a steep decline, followed by a long period of relatively low production. Output typically declines by between 50% and 75% in the first year of production, and most recoverable gas is usually extracted after just a few years.

### Well abandonment

At the end of their economic life, wells need to be safely abandoned, facilities **dismantled** and land returned to its natural state or put to new appropriate productive use. Long-term prevention of leaks to aquifers or to the surface is particularly important. Since much of the abandonment will not take place until production has ceased, the **regulatory** framework needs to ensure that the companies concerned make the necessary financial provisions and maintain technical capacity beyond the field's economic life to ensure that abandonment is completed satisfactorily, and well **integrity** maintained over the long term.

## 🔲 Words and Expressions

| | |
|---|---|
| injection | 注射;注入 |
| wellbore | 井筒,井眼 |
| carbonate | 碳酸盐 |
| fissure | 裂开;裂缝 |
| multiple | 多重的;多个的 |
| proppant | 支撑剂;压裂支撑剂 |
| premium | 保险费;额外费用 |
| venting | 排气 |
| flaring | 火炬骤燃 |
| dismantle | 拆卸;取消;解散 |
| regulatory | 管理的;控制的;调整的 |
| integrity | 完整;正直;诚实;廉政 |

## 🔲 Phrases and Expressions

| | |
|---|---|
| ceramic bead | 陶瓷球 |
| chemical additive | 化学添加剂 |
| prop open | 撑开 |
| multi-stage fracturing | 多段压裂 |
| single-stage fracturing | 一段压裂 |
| green completion | 绿色完井 |
| Christmas tree | 采油树 |
| reduced emissions completion | 减少排放完井 |

## 🔲 Language Focus

1. However, while collecting and processing the fluid is standard practice, capturing and selling the gas during this initial flow-back phase requires investment in gas separation and processing facilities, which does not always take place.

(参考译文:然而,尽管按常规应该收集和处理加工反排流体,但是在返排初期收集并出售气体需要投资配备气体分离和加工设施,这不是都能做到的。)

本句中 while 引导让步状语从句，相当于 though；which 引导非限制性定语从句，which 指代 investment in gas separation and processing facilities。

2. During production, wells will produce hydrocarbons and waste streams, which have to be managed.

（参考译文：油井产出的碳氢化合物和废液要设法处理。）

本句中 which 引导非限制性定语从句，which 指代"hydrocarbons and waste streams"。

3. Since much of the abandonment will not take place until production has ceased, the regulatory framework needs to ensure that the companies concerned make the necessary financial provisions and maintain technical capacity beyond the field's economic life to ensure that abandonment is completed satisfactorily, and well integrity maintained over the long term.

（参考译文：只有停产才会弃井，因此，有关规章制度应保证相关公司留有足够的财力和技术维护能力，在油田经济寿命结束后，还能保证弃井工作顺利完成，从长远角度维护井的完整性。)

本句为复合句，Since 引导的原因状语从句中含有 not...until... 结构，意为：直到……才……。主句中谓语 ensure 后面有个 that 引导的宾语从句，从句中主语 the companies 带了两个并列谓语：make... and maintain...；to ensure 为不定式短语做目的状语，后面的 that 宾语从句中含有 and 连接的两个并列句，well integrity 后省略了 is。

## Reinforced Learning

### Ⅰ. Answer the following questions for a comprehension of the text.

1. What does acid treatment refer to?

2. Which technique has played a key role in unlocking production of shale gas and light tight oil in theUnited States?

3. Why is gas vented or flared in shale and tight gas production?

4. How long might the production of a conventional well last?

5. What should be done after abandoning wells?

### Ⅱ. Multiple choice：choose the correct one from the alternative answers to give the exact meaning of the words.

1. Wells with long horizontal or lateral sections ( known as horizontal

wells) can increase <u>dramatically</u> the contact area between the reservoir rock and the wellbore.

    A. greatly      B. particularly      C. immediately      D. intangibly

    2. Hydraulic fracturing, developed <u>initially</u> in the late 1940s, is another effective and commonly-practised technology for low-permeability reservoirs.

    A. at last      B. at first      C. poorly      D. familiarly

    3. Once the pressure is <u>released</u>, these fractures would tend to close again and not produce any lasting improvement in the flow of hydrocarbons.

    A. increased      B. reduced      C. extended      D. discharged

    4. Some of the fluid <u>injected</u> during the process flows back up the well as part of the produced stream.

    A. instilled      B. fixed      C. derided      D. increased

    5. Whereby the hydrocarbons are separated from the fracturing fluid (and then sold) and the residual flow-back fluid is collected for processing and <u>recycling</u> or disposal.

    A. investing      B. reusing      C. conducting      D. correlating

    6. Venting and/or flaring of the gas at this stage are the main reasons why shale and tight gas can give rise to higher greenhouse-gas <u>emissions</u> than conventional production.

    A. objection      B. disservice      C. discharge      D. injection

    7. In some cases, the operator may decide to repeat the hydraulic fracturing <u>procedure</u> at later times in the life of the producing well.

    A. process      B. result

    C. forage      D. enhancement

    8. To ensure that the companies concerned make the necessary <u>financial provisions</u>.

    A. incomes      B. rules      C. funds      D. graft

    9. Output typically declines by between 50% and 75% in the first year of production, and most recoverable gas is usually <u>extracted</u> after just a few years.

    A. collected      B. grabbled      C. inducted      D. moved

    10. Long-term prevention of <u>leaks</u> to aquifers or to the surface is particularly important.

    A. draining      B. mortality      C. division      D. fusion

**III. Multiple choice：read the four suggested translations and choose the best answer.**

1. Acid treatment, involving the <u>injection</u> of small amounts of strong acids into the reservoir to dissolve some of the rock minerals.

A. 注入　　　B. 增加　　　C. 铺上　　　D. 倒掉

2. Advances in the <u>application</u> of these two techniques, in combination, largely explain the surge in shale gas production in the United States since 2005.

A. 类型　　　B. 应用　　　C. 条件　　　D. 发明

3. In many cases, a series of fractures is created at set <u>intervals</u>, one after the other, about every 100 metres along the horizontal well bore.

A. 歇息　　　B. 间隔　　　C. 停止　　　D. 连续

4. A standard single-stage hydraulic fracturing may pump down several hundredcubic metres of water together with <u>proppant</u> and a mixture of various chemical additives.

A. 催化剂　　　B. 抗酸剂　　　C. 增稠剂　　　D. 支撑剂

5. The repeated stresses on the well from multiple high-pressure procedures increase the <u>premium</u> on good well design and construction.

A. 投入　　　B. 运费　　　C. 价格　　　D. 方式

6. To ensure that gas bearing formations are completely <u>isolated</u> from other strata penetrated by the well.

A. 相容　　　B. 分散　　　C. 脱离　　　D. 集合

7. Whereby the hydrocarbons are <u>separated</u> from the fracturing fluid (and then sold) and the residual flow-back fluid is collected for processing and recycling or disposal.

A. 分离　　　B. 来源　　　C. 融合　　　D. 加入

8. In these cases, there can be <u>venting</u> of gas to the atmosphere (mostly methane, with a small fraction of VOCs).

A. 吸入　　　B. 创造　　　C. 挥发　　　D. 排放

9. At the end of their economic life, wells need to be safely abandoned, facilities <u>dismantled</u> and land returned to its natural state or put to new appropriate productive use.

A. 安装　　　B. 拆卸　　　C. 分裂　　　D. 组装

10. To ensure that abandonment is completed satisfactorily, and well integ-

rity maintained over the long term.

    A. 分散        B. 完整        C. 集中        D. 融合

### IV. Put the following sentences into Chinese.

1. During this flow-back period, typically over days (for a single-stage fracturing) to weeks (for a multi-stage fracturing), the amount of flow back of fracturing fluid decreases, while the hydrocarbon content of the produced stream increases, until the flow from the well is primarily hydrocarbons.

2. In many cases, a series of fractures is created at set intervals, one after the other, about every 100 metres along the horizontal well bore.

3. In some cases, the operator may decide to repeat the hydraulic fracturing procedure at later times in the life of the producing well, a procedure called re-fracturing.

4. Best practice during this period is to use a so-called "green completion" or "reduced emissions completion", whereby the hydrocarbons are separated from the fracturing fluid (and then sold) and the residual flow-back fluid is collected for processing and recycling or disposal.

5. At the end of their economic life, wells need to be safely abandoned, facilities dismantled and land returned to its natural state or put to new appropriate productive use.

### V. Put the following paragraphs into Chinese.

1. Several technologies have been developed over the years to enhance the flow from lowpermeability reservoirs. Acid treatment, involving the injection of small amounts of strong acids into the reservoir to dissolve some of the rock minerals and enhance the permeability of the rock near the wellbore, is probably the oldest and is still widely practised, particularly in carbonate reservoirs.

2. Hydraulic fracturing involves pumping a fluid-known as fracturing fluid-at high pressure into the well and then, far below the surface, into the surrounding target rock. This creates fractures or fissures a few millimeters wide in the rock. These fissures can extend tens or, in some cases, even hundreds of meters away from the well bore. Once the pressure is released, these fractures would tend to close again and not produce any lasting improvement in the flow of hydrocarbons. To keep the fractures open, small particles, such as sand or ceramic beads, are added to the pumped fluid to fill the fractures and to act as prop

pants, i. e. they prop open the fractures thus allowing the gas to escape into the well.

## 3.3　Coalbed Methane Development

### Guidance to Reading

*The term **coalbed methane** refers to methane adsorbed into the solid matrix of the coal, which can be extracted from coal beds. Some coal beds have long been known to be " gassy," and as a safety measure, boreholes were drilled into the seams from the surface, and the methane allowed to vent before mining. Coalbed methane as a natural-gas resource received a major push from the US federal government in the late 1970s. The US Department of Energy funded research into a number of unconventional gas sources, including coalbed methane. Coalbed methane was exempted from federal price controls, and was also given a federal tax credit. In Australia, commercial extraction of coal seam gas began in 1996.*

### Text

Coalbed methane refers to methane ( natural gas) held within the solid matrix of **coal seams**. Some of the methane is stored within the coal as a result of a process called **adsorption**, whereby a **film** of methane is created on the surface of the pores inside the coal. Open **fractures** in the coal may also contain free gas or water. In some cases, methane is present in large volumes in coalbeds and can constitute a serious safety hazard for coal-mining operations. Significant volumes of $CO_2$ may also be present in the coal.

There are both similarities and differences between coalbed methane and the two other main types of unconventional gas discussed, which are linked to the way in which coalbed methane is extracted, the associated costs and the impact on the environment. The main similarity is the low permeability of the gas-bearing reservoir-a critical factor for the technical and economic **viability** of extraction. Virtually all the permeability of a coalbed is due to fractures, in the form of **cleats and joints**. These fractures tend to occur naturally so that, within a small part of the seam, methane is able to flow through the coalbed. As with shale and tight gas deposits, there are major variations in the concentration of gas from one area to another within the coal seams. This, together with variations in the thickness of the seam, has a significant impact on potential production rates.

Above ground, coalbed methane production involves disruption to the landscape and local environment through the construction of drilling pads and access roads, and the **installation** of on-site production equipment, gas processing and transportation facilities. As is often the case with shale gas and tight gas, coalbed methane developments require the drilling of more wells than conventional oil and gas production; as a result, traffic and vehicle noise levels, noise from **compressors**, air pollution and the potential damage to local ecological systems are generally more of an issue than for conventional gas output.

There are some important differences between coalbed methane and shale or tight gas resources. Coalbed methane deposits can be located at shallow depths (these are predominantly the deposits that have been exploited thus far), whereas shale and tight gas are usually found further below the surface. Water is often present in the coalbed, which needs to be removed to allow the gas to flow to the well. In addition, coalbed methane contains very few heavier liquid hydrocarbons (natural gas liquids or **gas condensate**), which means the commercial viability of production depends heavily on the price at which the gas itself can be sold; in the case of shale gas produced together with large volumes of associated natural gas liquids, the price of oil plays a very important role in determining the overall **profitability** of the development project as shown in Figure 3.1.

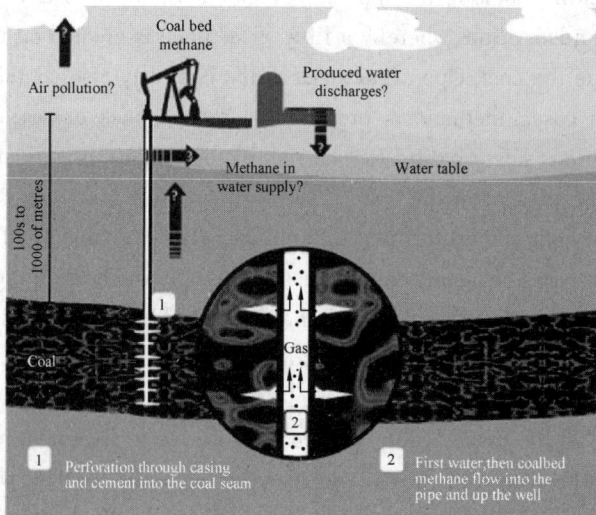

**Figure 3.1 Coalbed methane production techniques and possible environmental hazards**

Considerable progress has been made over the last 25 years in **honing techniques** to extract coalbed methane on a commercial basis, paving the way to production on a significant scale, initially in North America and, since the mid – 1990s, in Australia.

Coalbed methane can be produced from vertical or horizontal wells. The latter are becoming increasingly common, though less so than for shale gas. Generally, the thinner the coal seam and the greater the depth of the deposit, the more likely it is that a horizontal well will be drilled. Although a depth of 800 to 1 200 metres is typical, in some cases coalbed methane is located in shallow formations as little as 100 metres below the surface, making it more economical to drill a series of vertical wells, rather than a horizontal well with extended reach along the coal seam. For shallow deposits, wells can often be drilled using water-well drilling equipment, rather than rigs designed for conventional hydrocarbon extraction, with **commensurately** cheaper costs. For deeper formations (400 to 1 200 metres), both vertical and horizontal wells are used and **custom-built** small drilling rigs, capable of handling blow-out risks, have been developed.

Once a well is drilled, the water in the coalbed is extracted, either under natural pressure or by using mechanical pumping equipment-a process known as **dewatering**. As subsurface pressure drops with dewatering, the flow of natural gas previously held in place by water pressure increases initially as it is released from the natural fractures or **cleats** within the coalbed. The gas is separated from the water at the surface and is then compressed and injected into a gas-gathering pipeline for onward transportation.

As in the case of shale gas, the rate of production of coalbed methane is often significantly lower than that achieved in conventional gas reservoirs; it also tends to reach a peak quickly as water is extracted, before entering a period of decline as the well pressure drops further. A well's typical lifespan is between five and fifteen years, with maximum gas production often achieved after one to six months of water removal. In most cases, the low natural permeability of the coal seam means that gas can flow into the well from only a small segment of the coal seam-a characteristic shared with shale and tight gas. As a result, a relatively large number of wells are required over the area of the coalbed, especially if they are drilled vertically.

In some cases, it may also be necessary to use hydraulic fracturing to increase the permeability of the coal seam in order to stimulate the release of water and gas. This is normally practised only in deeper wells, typically at several hundred metres below the ground. The decision to proceed with hydraulic fracturing needs to be made before drilling begins, as the well and surface facilities need to be designed accordingly. The approach is similar to that described above, but in contrast to current practice with shale gas and tight gas wells, fracturing for coal-bed methane production is frequently a single-stage process, *i. e.* one fracturing job per well, rather than multi-stage. Since wells are often drilled **in batches**, the water required for hydraulic fracturing can be sourced from neighbouring wells that are being de-watered. The flow-back fluids recovered from the well are pumped to lined containment pits or tanks for treatment or **offsite** disposal.

## Words and Expressions

| | |
|---|---|
| adsorption | 吸附,吸附作用 |
| film | 薄膜 |
| fracture | 裂缝 |
| viability | 可行性,生存能力 |
| installation | 安装;装置 |
| compressors | 压缩机 |
| profitability | 盈利率;收益能力 |
| commensurately | 相当地;相称地 |
| dewatering | 气井排水;脱水作用 |
| cleat | 裂缝 |
| offsite | 厂区外,装置外 |

## Phrases and Expressions

| | |
|---|---|
| coalbed methane | 煤气层 |
| coal seams | 煤层 |
| cleats and joints | 节理 |
| on-site | 现场的 |
| gas condensate | 凝析油 |
| honing technique | 珩磨技术 |
| custom-built | 定制的 |

| in batches | 成批地；分批地 |

## Language Focus

1. There are both similarities and differences between coalbed methane and the two other main types of unconventional gas discussed, which are linked to the way in which coalbed methane is extracted, the associated costs and the impact on the environment.

（参考译文：煤层气同其他两种讨论过的主要非常规气既相似，也有差异，主要体现在开采方式、开采成本以及对环境的影响等方面。）

本句中主句为 there be 结构，"which are linked to..."是非限制性定语从句，which 在从句中做主语，指前面整句话的内容，即"煤层气同其他两种讨论过的主要非常规气既相似，也有差异，"；which are linked to 后有三个并列的介词宾语，即 the way，the associated costs 和 the impact；而 the way 后又有一个 in which 引导的定语从句，修饰先行词 the way，意为"in the way"。

2. Water is often present in the coalbed, which needs to be removed to allow the gas to flow to the well.

（参考译文：煤层中经常有水，因此必须排水才能使气体流入井中。）

本句中 which 引导非限制性定语从句，which 指代前句中的"water"。为了句子结构的平衡，先行词和定语从句分开了。

## Reinforced Learning

### I. Answer the following questions for a comprehension of the text.

1. This, together with variations in the thickness of the seam, has a significant impact on potential production rates. What does "this" refer to in this sentence?

2. In what way can coalbed methane production do harm to the environment?

3. What are the differences between coalbed methane and shale or tight gas resources?

4. Why is a relatively large number of wells required over the area of the coalbed, especially if they are drilled vertically?

5. Why does the decision to proceed with hydraulic fracturing need to be made before drilling begins?

Ⅱ. **Multiple choice: choose the correct one from the alternative answers to give the exact meaning of the words.**

1. In some cases, methane is present in large volumes in coalbeds and can constitute a serious safety hazard for coal-mining operations.

A. danger       B. muck       C. protection       D. decline

2. As with shale and tight gas deposits, there are major variations in the concentration of gas from one area to another within the coal seams.

A. patterns       B. increases       C. relations       D. differences

3. Above ground, coalbed methane production involves disruption to the landscape and local environment through the construction of drilling pads and access roads.

A. construction       B. damage       C. form       D. obscurity

4. The price of oil plays a very important role in determining the overall profitability of the development project.

A. partial       B. obsolete       C. entire       D. potent

5. There are both similarities and differences between coalbed methane and the two other main types of unconventional gas discussed.

A. normal       B. uncommon       C. repellent       D. precarious

6. Which means the commercial viability of production depends heavily on the price at which the gas itself can be sold.

A. puts. . . on       B. looks. . . on       C. views. . . on       D. relies. . . on

7. For shallow deposits, wells can often be drilled using water-well drilling equipment, rather than rigs designed for conventional hydrocarbon extraction, with commensurately cheaper costs.

A. quickly       B. mentally       C. considerably       D. visually

8. The gas is separated from the water at the surface and is then compressed and injected into a gas-gathering pipeline for onward transportation.

A. condensed       B. enlarged       C. relapsed       D. rended

9. A well's typical lifespan is between five and fifteen years, with maximum gas production often achieved after one to six months of water removal.

A. decoy       B. age       C. essence       D. lifetime

10. The decision to proceed with hydraulic fracturing needs to be made before drilling begins, as the well and surface facilities need to be designed accordingly.

A. particularly                B. correspondingly

C. specially                   D. bountifully

**Ⅲ. Multiple choice：read the four suggested translations and choose the best answer.**

1. The low natural permeability of the coal seam means that gas can flow into the well from only a small <u>segment</u> of the coal seam.

　　A. 部分　　　B. 间隔　　　C. 章节　　　　D. 整体

2. It may also be necessary to use hydraulic fracturing to increase the permeability of the coal seam in order to <u>stimulate</u> the release of water and gas.

　　A. 反响　　　B. 刺激　　　C. 回味　　　　D. 释放

3. The main similarity is the low permeability of the gas-bearing reservoir-a critical factor for the technical and economic <u>viability</u> of extraction.

　　A. 安全性　　B. 有效性　　C. 可行性　　　D. 稳定性

4. For shallow deposits，wells can often be drilled using water-well drilling equipment，rather than rigs designed for conventional hydrocarbon <u>extraction</u>.

　　A. 处理　　　B. 变质　　　C. 反应　　　　D. 提取

5. Both vertical and horizontal wells are used and <u>custom-built</u> small drilling rigs，capable of handling blow-out risks，have been developed.

　　A. 成型的　　B. 定制的　　C. 建设的　　　D. 服务的

6. It also tends to reach a <u>peak</u> quickly as water is extracted，before entering a period of decline as the well pressure drops further.

　　A. 顶峰　　　B. 山谷　　　C. 阶段　　　　D. 山峦

7. A well's typical lifespan is between five and fifteen years，with maximum gas production often achieved after one to six months of <u>water removal</u>.

　　A. 流动　　　B. 脱水　　　C. 洗涤　　　　D. 水流

8. In some cases，it may also be necessary to use <u>hydraulic</u> fracturing to increase the permeability of the coal seam in order to stimulate the release of water and gas.

　　A. 水力的　　B. 排水的　　C. 液压的　　　D. 积压的

9. The approach is similar to that described above，but <u>in contrast to</u> current practice with shale gas and tight gas wells.

　　A. 与……反应　　　　B. 与……相符

　　C. 与……混合　　　　D. 与……形成对比

10. Since wells are often drilled <u>in batches</u>，the water required for hydraulic fracturing can be sourced from neighbouring wells that are being de-watered.

A. 分批地　　　B. 一个一个地　C. 逐渐地　　　　D. 一齐地

## IV. Put the following sentences into Chinese.

1. Some of the methane is stored within the coal as a result of a process called adsorption, whereby a film of methane is created on the surface of the pores inside the coal.

2. In some cases, methane is present in large volumes in coalbeds and can constitute a serious safety hazard for coal-mining operations.

3. Considerable progress has been made over the last 25 years in honing techniques to extract coalbed methane on a commercial basis, paving the way to production on a significant scale, initially in North America and, since the mid – 1990s, in Australia.

4. The gas is separated from the water at the surface and is then compressed and injected into a gas-gathering pipeline for onward transportation.

5. The flow-back fluids recovered from the well are pumped to lined containment pits or tanks for treatment or offsite disposal.

## V. Put the following paragraphs into Chinese.

1. Above ground, coalbed methane production involves disruption to the landscape and local environment through the construction of drilling pads and access roads, and the **installation** of on-site production equipment, gas processing and transportation facilities. As is often the case with shale gas and tight gas, coalbed methane developments require the drilling of more wells than conventional oil and gas production; as a result, traffic and vehicle noise levels, noise from compressors, air pollution and the potential damage to local ecological systems are generally more of an issue than for conventional gas output.

2. There are some important differences between coalbed methane and shale or tight gas resources. Coalbed methane deposits can be located at shallow depths, whereas shale and tight gas are usually found further below the surface. Water is often present in the coalbed, which needs to be removed to allow the gas to flow to the well. In addition, coalbed methane contains very few heavier liquid hydrocarbons, which means the commercial viability of production depends heavily on the price at which the gas itself can be sold; in the case of shale gas produced together with large volumes of associated natural gas liquids, the price of oil plays a very important role in determining the overall profitability of the devel-

opment project.

# 3.4 Coalbed Methane Production and Effects on Groundwater

## Guidance to Reading

As in the case of shale gas, the rate of production of coalbed methane is often significantly lower than that achieved in conventional gas reservoirs. In some cases, it may also be necessary to use hydraulic fracturing to increase the permeability of the coal seam in order to stimulate the release of water and gas. However, since the productive coal seams are often at shallower depths than tight or shale gas deposits, there is a greater risk that fracturing fluids might enter an aquifer. Sometimes the well fluids may leak into the aquifer behind the well casing, and the methane might also leak through the rock from the producing zone. As with shale or tight gas production, the flow-back fluids removed from the well after fracturing need to be treated before disposal, but the waste water is occasionally insufficiently treated and disposed into the groundwater.

## Text

There are concerns about the impact of coalbed methane production on groundwater flows and the supply and purity of water in aquifers **adjacent to** the coal seams being exploited. The extent to which this can occur is very location specific and depends on several factors, the most important of which are the overall volume of water initially in the coalbed and the **hydrogeology** of the **basin**; the density of the coalbed methane wells; the rate of water pumping by the operator; the **connectivity** of the coalbed and aquifer to surrounding water sources and, therefore, the rate of **recharge** of the aquifer; and the length of time over which pumping takes place.

In theUnited States, various agencies now monitor water in producing areas in order to learn more about this process. Depletion of aquifers because of coalbed methane production has been **well-documented** in the Powder River Basin: in the Montana portion of the basin, 65% to 87% recovery of coalbed groundwater levels has occurred after production ceased. However, the extent to which water levels in shallow **alluvial** and water table aquifers have dropped has not been measured (recent legislation in Queensland in Australia now requires such **measurements** to be performed). There is evidence that groundwater movement provoked by dewatering during coalbed methane production has increased the amount of dissolved salt and other minerals in some areas.

Because productive coal seams are often at shallower depths than tight or shale gas deposits, there is also a greater risk that fracturing fluids might **find their way into** an aquifer directly or via a fracture system (either a natural system or one that is created through fracturing). This risk is **mitigated** in part by the fact that, in contrast to shale or tight gas, the dewatering required for production of coalbed methane means that less water may be left in the ground in aquifers **near the vicinity of** the well, limiting the potential for contamination. As with shale or tight gas production, the flow-back fluids removed from the well after fracturing need to be treated before disposal.

The first hazard-the risk of spills at the surface-can be mitigated through rigorous containment of all fluid and solid streams. Accidents can always happen but good procedures, training of personnel and availability of spill control equipment can ensure they have a limited impact. Greater use of pipelines to move liquids can reduce the risks associated with trucking movements.

Controlling the second hazard-**leakage** into a shallow aquifer behind the well casing-requires use of best practice in well design and well construction, particularly during the **cementing** process, to ensure a proper seal is in place, systematic verification of the quality of the seal and ensuring the seal does not **deteriorate** through the life of a well. This is a particular issue for wells in which multi-stage hydraulic fracturing is performed: the repeated cycles of high pressure pumping can apply repeated stress to the casing and to the **cement column**, potentially weakening them; selection of an appropriate strength of casing is therefore important.

The third hazard-leakage through the rock from the producing zone-is unlikely in the case of shale gas or tight gas because the producing zone is one to several thousand metres below any relevant aquifers and this thickness of rock usually includes one or several very **impermeable layers**. For example, the deepest potential underground sources of drinking water in the Barnett shale are at a depth of 350 metres, whereas the shale layer is at 2 000 to 2 300 metres. However, the hazard may be **encountered** if the producing zone is shallower or if there are shallow pockets of naturally occurring methane above the target reservoir. It is also theoretically possible if there are no identified impermeable layers in between or if deep **faults** are present that can act as a **conduit** for fluids to move from the deep producing zone towards the surface (such fluid movements are generally slow, but can occur on **time scales** of tens of years). One particular possibility is that hydraulic fractures may not be contained in the targeted rock layer and may break through important rock barriers or connect to deep faults. This is a rare occurrence because hydraulic fracturing is designed to avoid this situation, but it cannot be completely excluded when the local geology is insufficiently understood.

Appropriate prior studies of the local geology to identify such situations are therefore a must before undertaking significant developments. Indeed, methane **seeps** to the surface have long been known (for example, the flame that has been burning for centuries in the village of Mrapen in Central Java, Indonesia, or the gas that fuels the "Eternal Flame Falls" in New York State, United States) and they have been used as a way to identify the presence of hydrocarbon deposits underground, showing that perfect rock seals do not always exist.

On the other hand, the existence of seeps, and for that matter the presence of methane in many aquifers, shows that not all contamination is linked to industrial activity; it can also occur as a result of natural geological or biological processes. Addressing the fourth hazard-discharge of insufficiently treated waste water into groundwater or, even, deep underground-requires a regulatory response including appropriate tracking and documentation of waste water volumes and composition, how they are transported and disposed.

## Words and Expressions

| | |
|---|---|
| hydrogeology | 水文地质学 |
| basin | 盆地 |
| connectivity | 连贯性 |
| recharge | 再袭击；再充电 |
| well-documented | 证据充足的 |
| alluvial | 冲击的 |
| measurement | 测量；尺寸 |
| mitigate | 镇静；缓和；减轻 |
| leakage | 泄露 |

| | |
|---|---|
| cementing | 胶接;粘合;注水泥 |
| deteriorate | 恶化 |
| encounter | 遭遇;遇到 |
| faults | 断层 |
| conduit | 导管;导水管 |
| seep | 渗出;渗流 |

## Phrases and Expressions

| | |
|---|---|
| adjacent to | 邻近的 |
| find their way into | 进入 |
| near the vicinity of | 在……附近 |
| cement column | 水泥环 |
| impermeable layer | 不透水层;防水层 |
| time scale | 时间尺度;时标 |

## Language Focus

1. Because productive coal seams are often at shallower depths than tight or shale gas deposits, there is also a greater risk that fracturing fluids might find their way into an aquifer directly or via a fracture system ( either a natural system or one that is created through fracturing).

[参考译文:由于煤层气产层往往比致密气藏和页岩气藏浅,因而压裂液直接或间接通过裂隙(自然形成的裂隙或经压裂形成的裂隙)进入含水层的风险也更大。]

本句为复合句,主句是 there be 结构,主语 risk 后是 that 引导的同位语从句,具体说明更大的风险是什么;because 引导原因状语从句;另外,括号里的 that 从句是定语从句修饰先行词 one,one 指代 fracture system。

2. This risk is mitigated in part by the fact that, in contrast to shale or tight gas, the dewatering required for production of coalbed methane means that less water may be left in the ground in aquifers near the vicinity of the well, limiting the potential for contamination.

(参考译文:跟页岩气和致密气开采比起来,由于开采煤层气需要进行气井排水,所以气井附近含水层里剩下的水会少一些,在一定程度上减少了水污染的可能性。)

本句中 fact 后 that 引导的是同位语从句,说明 fact 的内容,这个同位语从句又含有一个 that 引导的宾语从句 that less water may be..., 做 means 的宾语。注意,同位语从句中的 dewatering 是主语,required for... 是过去分词短语作后置定语,修饰 dewatering; limiting the potential for contamination 是现在分词短语做结果状语。

## 🔲 Reinforced Learning

### I. Answer the following questions for a comprehension of the text.

1. What are people's concerns over the impact of coalbed methane production?

2. Why is there a greater risk that fracturing fluids might enter an aquifer directly or via a fracture system in coalbed methane production?

3. Why is it important to select an appropriate strength of casing during the well completion?

4. Why does the author say "the third hazard-leakage through the rock from the producing zone-is unlikely in the case of shale gas or tight gas"?

5. What can be done to address the fourth hazard-discharge of insufficiently treated waste water into groundwater or, even, deep underground?

### II. Multiple choice: choose the correct one from the alternative answers to give the exact meaning of the words.

1. There are concerns about the impact of coalbed methane production on groundwater flows and the supply and purity of water in aquifers adjacent to the coal seams being exploited.

    A. next to     B. far away from     C. in addition to     D. by means of

2. In the Montana portion of the basin, 65% to 87% recovery of coalbed groundwater levels has occurred after production ceased.

    A. part       B. integrity      C. entity      D. efficacy

3. Depletion of aquifers because of coalbed methane production has been well-documented in the Powder River Basin.

    A. damage     B. configuration     C. exhaustion     D. protection

4. There is evidence that groundwater movement provoked by dewatering during coalbed methane production has increased the amount of dissolved salt and other minerals in some areas.

A. built      B. betrayed      C. aroused      D. concur

5. There are concerns about the impact of coalbed methane production on groundwater flows and the supply and purity of water in aquifers adjacent to the coal seams being underlined{exploited}.

     A. developed    B. confided      C. invested      D. made use of

6. This risk is mitigated in part by the fact that, in contrast to shale or tight gas, the dewatering required for production of coalbed methane means that less water may be left in the ground.

     A. aggravated    B. alleviated      C. ruined      D. broken

7. The first hazard-the risk of spills at the surface-can be mitigated through rigorous containment of all fluid and solid streams.

     A. fragments    B. leaks      C. pieces      D. pebbles

8. Systematic verification of the quality of the seal and ensuring the seal does not deteriorate through the life of a well.

     A. investigate    B. utilize      C. append      D. worsen

9. This is a rare occurrence because hydraulic fracturing is designed to avoid this (potentially costly) situation, but it cannot be completely excluded when the local geology is insufficiently understood.

     A. well      B. not enough      C. definitely      D. categorically

10. Appropriate prior studies of the local geology to identify such situations are therefore a must before undertaking significant developments.

     A. previous    B. elaborative      C. chaste      D. thorough

**III. Multiple choice: read the four suggested translations and choose the best answer.**

1. However, the extent to which water levels in shallow alluvial and water table aquifers have dropped has not been measured (recent legislation in Queensland in Australia now requires such measurements to be performed).

     A. 品格      B. 价值      C. 规格      D. 测量

2. There is evidence that groundwater movement provoked by dewatering during coalbed methane production has increased the amount of dissolved salt and other minerals in some areas.

     A. 暗示      B. 解决      C. 分散      D. 溶解

3. The dewatering required for production of coalbed methane means that less water may be left in the ground in aquifers near the vicinity of the well,

limiting the potential for contamination.

    A. 临近        B. 距离        C 观察        D. 调研

4. The first hazard-the risk of spills at the surface-can be mitigated through rigorous containment of all fluid and solid streams.

    A. 严格的        B. 辉煌的        C. 无法修复的    D. 激烈的

5. Accidents can always happen but good procedures, training of personnel and availability of spill control equipment can ensure they have a limited impact.

    A. 获得的可能性    B. 建设性        C. 完整        D. 效度

6. Systematic verification of the quality of the seal and ensuring the seal does not deteriorate through the life of a well.

    A. 核实        B. 确信        C. 准确        D. 损坏

7. If deep faults are present that can act as a conduit for fluids to move from the deep producing zone towards the surface.

    A. 断层        B. 错误        C. 迷失        D. 爆破

8. This is a rare occurrence because hydraulic fracturing is designed to avoid this (potentially costly) situation, but it cannot be completely excluded.

    A. 核实        B. 考虑        C. 包括        D. 排除

9. The third hazard is unlikely in the case of shale gas or tight gas because the producing zone is one to several thousand metres below any relevant aquifers and this thickness of rock usually includes one or several very impermeable layers.

    A. 透明层        B. 地质层        C. 防水层        D. 页岩层

10. Appropriate prior studies of the local geology to identify such situations are therefore a must before undertaking significant developments. Indeed, methane seeps to the surface have long been known.

    A. 埋藏        B. 渗出        C. 保存        D. 流动

**IV. Put the following sentences into Chinese.**

1. There are concerns about the impact of coalbed methane production on groundwater flows and the supply and purity of water in aquifers adjacent to the coal seams being exploited.

2. There is evidence that groundwater movement provoked by dewatering during coalbed methane production has increased the amount of dissolved salt and other minerals in some areas.

3. As with shale or tight gas production, the flow-back fluids removed from the well after fracturing need to be treated before disposal.

4. It is also theoretically possible if there are no identified impermeable layers in between or if deep faults are present that can act as a conduit for fluids to move from the deep producing zone towards the surface.

5. Appropriate prior studies of the local geology to identify such situations are therefore a must before undertaking significant developments.

### V. Put the following paragraphs into Chinese.

1. Because productive coal seams are often at shallower depths than tight or shale gas deposits, there is also a greater risk that fracturing fluids might find their way into an aquifer directly or via a fracture system. This risk is mitigated in part by the fact that, in contrast to shale or tight gas, the dewatering required for production of coalbed methane means that less water may be left in the ground in aquifers near the vicinity of the well, limiting the potential for contamination. As with shale or tight gas production, the flow-back fluids removed from the well after fracturing need to be treated before disposal.

2. One particular possibility is that hydraulic fractures may not be contained in the targeted rock layer and may break through important rock barriers or connect to deep faults. This is a rare occurrence because hydraulic fracturing is designed to avoid this (potentially costly) situation, but it cannot be completely excluded when the local geology is insufficiently understood.

# Chapter 4　Unconventional Oil and Gas Production Technologies

## 4. 1　Exploration

### ⌗ Guidance to Reading

*Seismic surveys are the methods most widely used in the exploration for oil and gas, providing more detailed information about the shape and depth of subsurface structures than any of the other geophysical methods, such as electric, magnetic, and gravity surveys. Using a special type of seismograph, or geophone, seismic surveys explore the geological structure in the earth's sedimentary section by recording the ground movements produced by manmade explosions or non-explosive seismic technology. In reflection seismology, the end product of seismic data processing is a seismic section that is analogous to a geologic cross section. Reflecting horizons are visible on the seismic sections and the interpreter can use the sections to map subsurface geologic structure.*

### ⌗ Text

Exploration is a critical initial process step in the development of oil and gas resources. Before shale gas can be extracted from deep beneath the earth's surface through drilling, **stimulation**, and extraction processes, precise estimates of the location and quantity of potential gas reserves and the geologic and geophysical properties of the subsurface must be developed to maximize the economic viability of subsequent steps and eventual gas production. E&P companies invest heavily in exploration technologies to maximize production and to minimize the risk and cost associated with drilling in non-productive locations.

The shale gas exploration process typically begins with geologists examining the surface structure of the earth and determining areas where it is likely that gas deposits might exist. After identifying an area where it is geologically possible for a natural gas formation to exist, various advanced technologies are used to develop an accurate mapping of underground formations. These include the use of seismic, magnetic, and gravity surveys.

### 2 – D Reflection Seismology

An effective and commonly applied technology **leverages** basic seismology, as shown in Figure 4. 1. Seismology is the study of the movement of seismic waves through the earth's crust and their varied interaction with different types of underground formations. Reflection seismology-based seismic surveys are commonly used to map the subsurface structure of rock formations.

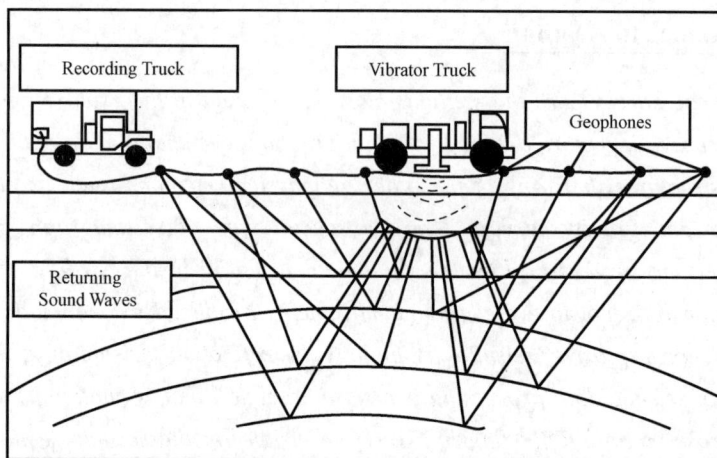

**Figure 4. 1　Visualization of Reflection Seismology（2 – D）Process**

Reflection seismology involves the generation of artificial seismic waves and subsequently capturing the subsurface reflections of these waves to create **multidimensional** representations of the underground geology. In the past, seismic waves were created using dynamite, however most companies now rely on non-explosive seismic technology. These technologies usually consist of wheeled or tracked vehicles which carry specialized equipment designed to create a series of vibrations. These in turn creates seismic waves which propagate underground and reflect off of the various subsurface layers. The seismic reflections are then picked up by sensitive geophones embedded or placed on the ground surface. The data received by the geophones are then transmitted to a seismic recording truck that records the data for further interpretation by geophysicists and petroleum resource engineers.

Until the 1980's, reflection seismic **acquisition** was carried out by arranging the source and receivers in a line then advancing the equipment along a linear transit to complete the survey, resulting in a two-dimensional (2 – D) seismic profile of a subsurface **cross-section**. 2 – D seismic surveys were extensively used in the oil and gas industry to map conventional resources for many decades. However, the success rate of such 2 – D profiles in identifying productive well locations was limited to around 50 percent.

### 3 – D Seismic Imaging

As E&P companies began to explore for more remote and deeper conventional and unconventional resources, the importance of accuracy in sub-surface surveying increased largely due to the higher risk and costs associated with the development of these resources. Three-dimensional (3 – D) seismic surveys, which leverage the same principles of 2 – D seismology but use significantly larger numbers of geophones arranged in a grid pattern, generate a more **granular** and 3 – D view of the subsurface. 3 – D seismology began to see increased adoption in the 1990s as the substantial increase in the probability of accurately locating productive areas more than offset the higher costs of 3 – D surveys.

Recent advances in acquisition technology have improved the quality and economics of 3 – D survey acquisition. In on-shore exploration applications, the logistics of **deploying** geophone receiver stations and cables are a significant expense and can severely limit the number of receivers that can be deployed. Using advanced electronics, the receiver signal is now commonly digitized and the data can be recorded locally or transmitted to a base station by radio or by **fiber-optic** cables. This improvement in data handling makes it economical to deploy more receivers during a survey. Additionally, improvements in seismic data processing and imaging capabilities have also improved the accuracy of predictive results. In fact, using 3 – D seismic techniques has been estimated to increase the likelihood of successful reservoir location by 50 percent over that achieved by 2 – D methods.

These advances combined with improvements in computational power over the past two decades have made the acquisition and processing of 3 – D seismic surveys even more feasible and economical. As a result, the vast majority of seismic data are now collected as 3 – D surveys, as shown in Figure 4. 2.

### 4 – D Seismic Imaging

A fairly recent development is the use of 4 – D seismic technologies. 4 – D seismology expands on 3 – D by adding time as the fourth dimension, allowing ob-

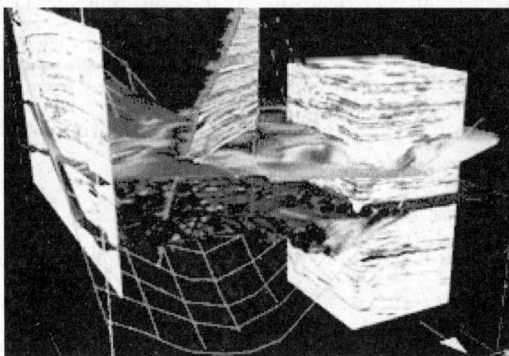

**Figure 4. 2    Advanced 3 – D Seismic Image**

servation of changes in subsurface characteristics over time. Since time is the fourth dimension, this technique is also called **4 – D time-lapse imaging**. Essentially, 4 – D seismic readings are taken sequentially over time to create a moving model of the sub-strata. A better understanding of properties such as underground fluid flow, **viscosity**, temperature, and **saturation** can be gained through analysis of these changes over time.

Using 4 – D surveys, finding costs are reduced as exploration teams can more readily identify natural gas prospects, place wells more effectively, reduce drilling costs, and cut exploration time. The technology enables a reduction in the number of dry holes drilled, translating to further economic and environmental benefits. Further, more accurate predictions of future reservoir production have demonstrated a major impact on ultimate gas recovery. In fact, recovery rates of 65 to 70 percent can result through the use of 4 – D imaging compared to 50 percent with 3 – D seismic imaging.

Advances in 3 – D and 4 – D reflection seismic **profiling** have enabled the industry to successfully identify and grow potential and proved reserves of shale and other unconventional gas over the past decade. These advanced exploration technologies are often cited along with advances in horizontal drilling and hydraulic fracturing as the key technological developments driving the economic viability of shale gas production. Nonetheless, seismic surveys are often complemented with data from magnetic and gravity surveys and actual physical data **garnered** from exploratory wells.

In addition to using seismology, the magnetic properties of underground formations can be measured to generate geological and geophysical data. This is accomplished through the use of **magnetometers**, which are devices to measure variations in the earth's magnetic field resulting from the presence of varying subsurface formations. Further, geophysicists can measure and record the minute differences in the earth's gravitational field to gain a better understanding of subsurface characteristics. By measuring these differences, geophysicists are able to evaluate the nature of underground formations and determine their potential for recoverable shale gas.

## Words and Expressions

| | |
|---|---|
| geophone | 检波器 |
| seismology | 地震学 |
| stimulation | 增产措施 |
| leverage | 手段, 影响力; 利用 |
| multi-dimensional | 多维的 |
| acquisition | 获得, 接收; 采集 |
| cross-section | 横截面 |
| granular | 颗粒的 |
| deploy | 展开, 配置, 部署 |
| fiber-optic | 光纤 |
| viscosity | 粘稠, 粘滞 |
| saturation | 饱和, 饱和状态 |
| profiling | 剖面法 |
| garner | 贮藏, 积累, 得到 |
| magnetometers | 磁力仪 |

## Phrases and Expressions

| | |
|---|---|
| seismic survey | 地震勘探 |
| magnetic survey | 磁法勘探 |
| gravity survey | 重力勘探 |
| reflection seismology | 反射地震学 |
| 2 – D reflection seismology | 二维反射地震学 |
| 3 – D seisimic imaging | 三维地震成像 |
| 4 – D time-lapse imaging | 四维延时成像 |
| carry out | 执行; 进行(实验、维修等) |

## ⊡ Language Focus

1. The shale gas exploration process typically begins with geologists examining the surface structure of the earth and determining areas where it is likely that gas deposits might exist.

（参考译文:页岩气勘探一般是地质学家先勘查地表结构,确定气藏可能存在的区域。）

本句中,with 短语是独立主格结构,geologists examining 是逻辑上的主谓结构,这个结构中含有一个 where 引导的定语从句,修饰先行词 areas,where 从句中还有一个 that 引导的主语从句,it 为形式主语。

2. Seismology is the study of the movement of seismic waves through the earth's crust and their varied interaction with different types of underground formations.

（参考译文:地震学研究的是地震波穿过地壳的运动,以及震波在不同类型地层中的相互作用。）

本句中 the movement of seismic waves through the earth's crust and their varied interaction with different types of underground formations 作 the study 的同位语,说明 the study 的具体内容,来定义 seismology。

3. The data received by the geophones are then transmitted to a seismic recording truck that records the data for further interpretation by geophysicists and petroleum resource engineers.

（参考译文:检波器将接收到的数据随后传送给地震波记录车存储,留待地球物理学家和石油资源工程师对其作进一步解释。）

本句中 received by the geophones 是过去分词短语作 the data 的后置定语,that records the data for further interpretation by geophysicists and petroleum resource engineers 是后置定语从句修饰其先行词 seismic recording truck。

4. This is accomplished through the use of magnetometers, which are devices to measure variations in the earth's magnetic field resulting from the presence of varying subsurface formations.

［参考译文:这就需要用磁力仪来完成了,即用来测量地球磁场各种变化(因地层多种多样)的仪器。］

本句含有一个 which 引导的非限制性定语从句,修饰前面的 magnetometers,从句中 to measure 是不定式短语作后置定语,修饰 devices,还有一个现在分词短语 resulting from … 做后置定语,修饰 variations。

## Reinforced Learning

**Ⅰ. Answer the following questions for a comprehension of the text.**

1. Give the definition of exploration.

2. What does seismology study?

3. What kinds of advanced technologies can be used to develop an accurate mapping of underground formations?

4. What benefit do advances in 3 – D and 4 – D reflection seismic profiling offer in petroleum exploration?

5. What are the advantages of 4 – D seismic technologies compared to 3 – D?

**Ⅱ. Multiple choice:choose the correct one from the alternative answers to give the exact meaning of the words.**

1. Precise estimates of the location and quantity of potential gas reserves and the geologic and geophysical properties of the subsurface must be developed to maximize the economic viability of subsequent steps and eventual gas production.

A. ability      B. feasibility      C. capability      D. practicability

2. Reflection seismology involves the generation of artificial seismic waves and subsequently capturing the subsurface reflections of these waves to create multi-dimensional representations of the underground geology.

A. relatives      B. reactions      C. refractions      D . reflexion

3. The data received by the geophones are then transmitted to a seismic recording truck that records the data for further interpretation by geophysicists and petroleum resource engineers.

A. explanation   B. interaction      C. instruction      D. information

4. The data received by the geophones are then transmitted to a seismic recording truck that records the data for further interpretation by geophysicists and petroleum resource engineers.

A. transferred   B. spread      C. transported      D . transformed

5. Three-dimensional (3 – D) seismic surveys, which leverage the same principles of 2 – D seismology but use significantly larger numbers of geophones arranged in a grid pattern, generate a more granular and 3 – D view of the sub-

surface.

    A. use        B. make        C . exploit        D. explore

    6. 3 – D seismology began to see increased adoption in the 1990s as the substantial increase in the probability of accurately locating productive areas more than offset the higher costs of 3 – D surveys.

    A. large        B. physical        C. significant        D. material

    7. Recent advances in acquisition technology have improved the quality and economics of 3 – D survey acquisition.

    A. acquaintance B. reception        C. attainment        D. possession

    8. A fairly recent development is the use of 4 – D seismic technologies. 4 – D seismology expands on 3 – D by adding time as the fourth dimension, allowing observation of changes in subsurface characteristics over time.

    A. feature        B. type        C. personality        D. chance

    9. Further, more accurate predictions of future reservoir production have demonstrated a major impact on ultimate gas recovery.

    A. endless        B. extreme        C. limited        D. final

    10. This is accomplished through the use of magnetometers, which are devices which measure variations in the earth's magnetic field resulting from the presence of varying subsurface formations.

    A. constitutions B. figurations        C. modalities        D. institutions

**III. Multiple choice : read the four suggested translations and choose the best answer.**

    1. Reflection seismology-based seismic surveys are commonly used to map the subsurface structure of rock formations.

    A. 底层的        B. 表层的        C. 潜伏的        D. 地下的

    2. Reflection seismology involves the generation of artificial seismic waves and subsequently capturing the subsurface reflections of these waves to create multi-dimensional representations of the underground geology.

    A. 因此        B. 随后        C. 所以        D. 然而

    3. These technologies usually consist of wheeled or tracked vehicles which carry specialized equipment designed to create a series of vibrations.

    A. 专业的        B. 特别的        C. 分类的        D. 多样的

    4. Additionally, improvements in seismic data processing and imaging capabilities have also improved the accuracy of predictive results.

A. 模糊的　　　B. 未知的　　　C. 猜测的　　　D. 预言性的

5. These advanced exploration technologies are often cited along with advances in <u>horizontal</u> drilling and hydraulic fracturing as the key technological developments driving the economic viability of shale gas production.

A. 立体的　　　B. 竖直的　　　C. 水平的　　　D. 平行的

**Ⅳ. Put the following sentences into Chinese.**

1. Reflection seismology involves the generation of artificial seismic waves and subsequently capturing the subsurface reflections of these waves to create multi-dimensional representations of the underground geology.

2. The data received by the geophones are then transmitted to a seismic recording truck that records the data for further interpretation by geophysicists and petroleum resource engineers.

3. Using advanced electronics, the receiver signal is now commonly digitized at or near the receiver and the data can be recorded locally or transmitted to a base station by radio or by fiber-optic cables.

4. Advances in 3 – D and 4 – D reflection seismic profiling have enabled the industry to successfully identify and grow potential and proved reserves of shale and other unconventional gas over the past decade.

5. 4 – D seismology expands on 3 – D by adding time as the fourth dimension, allowing observation of changes in subsurface characteristics over time.

**Ⅴ. Put the following paragraphs into Chinese**

1. Exploration is a critical initial process step in the development of oil and gas resources. Before shale gas can be extracted from deep beneath the earth's surface through drilling, stimulation, and extraction processes, precise estimates of the location and quantity of potential gas reserves and the geologic and geophysical properties of the subsurface must be developed to maximize the economic viability of subsequent steps and eventual gas production.

2. Three-dimensional (3 – D) seismic surveys, which leverage the same principles of 2 – D seismology but use significantly larger numbers of geophones arranged in a grid pattern, generate a moregranular and 3 – D view of the subsurface. 3 – D seismology began to see increased adoption in the 1990s as the substantial increase in the probability of accurately locating productive areas more than offset the higher costs of 3 – D surveys.

## 4.2 Well Drilling

### 🔲 Guidance to Reading

*Wells are generally classified as two kinds: exploration wells and **development wells**. An exploration well, or 'wildcat', is one that is drilled primarily for the purpose of determining that oil or gas actually exists in a subsurface rock formation. It is only after a well is drilled into the formation that the presence of oil or gas can definitely be confirmed or denied to be present. Drilling can be done vertically (vertical well) or at an angle (directional drilling) which can hit targets to avoid certain topographical features in ways that cannot be achieved with a vertical well. Directional and horizontal drilling has been used to reach targets beneath adjacent lands, reduce the footprint of gas field development, and increase the length of the "pay zone" in a well.*

### 🔲 Text

#### Exploratory Wells

Despite advances in remote and surface-based exploration technologies, a more comprehensive understanding of subsurface geology and **verification** of the presence and recoverability of shale gas requires the drilling of an exploratory well. Geologists can examine the drill cuttings and fluids to gain a better understanding of the geologic features of the subsurface area. Formation evaluation, or well logging, is a commonly used technique in development as well as **exploratory wells**.

A number of exploratory wells may need to be analyzed in order to sufficiently characterize the potential of ashale formation, particularly if the geologic basin is large and there are variations in the target shale zone. Since drilling operations are **time-consuming** and expensive, exploratory wells are only drilled in areas where other data has indicated a high probability of accessible shale gas formations.

#### Well Pad Construction

The first step in developing a shale gas well site is construction of access roads and the well pad. A well pad is a surface location used for locating the rig during drilling and producing from once the well is completed. Roads provide access to the site for well construction and **maintenance**. Wells drilled vertically require one well pad and related access road to be constructed per well, while one well pad and access road can support multiple horizontally drilled wells.

Constructing access roads involves clearing the route and preparing the surface for movementof heavy equipment. Once cleared, the ground surface is covered with a layer of gravel and road **sedimentation**, **erosion**, and **drainage** control features such as **drainage ditches** and **culverts** are constructed. The size of the access road is dictated by the distance of the well pad from an existing road, the route the access road is required to take to avoid certain **topographical** features, and the size of equipment to be transported to the well site during construction and drilling.

Site preparation for constructing the well pad primarily involves clearing and leveling an appropriately sized area to support the drilling equipment footprint and related containment pits for drilling fluid and water and preparing the surface to support the movement of heavy equipment. Ground surface preparation typically involves placing a layer of gravel over the cleared area, establishment of sediment and erosion control features for the well pad itself, and construction of pits to contain drilling fluid and water. Additionally, **gas gathering systems** are constructed to connect individual wells to pipelines for transporting gas.

The surface area required for the well pad itself, **exclusive of** utility access and roads, ranges from two to five acres during drilling and fracturing to allow room for equipment movement and **staging** and containment pits. During gas production, the area no longer required for these purposes is typically **reclaimed**, which reduces the area of the actual well pad to one to three acres, as shown in Figure 4. 3.

### *Well Drilling*

Both vertical and horizontal wells have been drilled to access shale gas. As technological advances have improved their economics and operational feasibility, horizontal wells have become more favored by the industry for many shale plays as these wells enable increased pay zone contact. Similar technology and equipment are employed for drilling both vertical and horizontal wells, however there are some differences. For instance, drilling a horizontal well requires the use of larger drilling rigs and the selection of types of drilling fluid used. However, the key

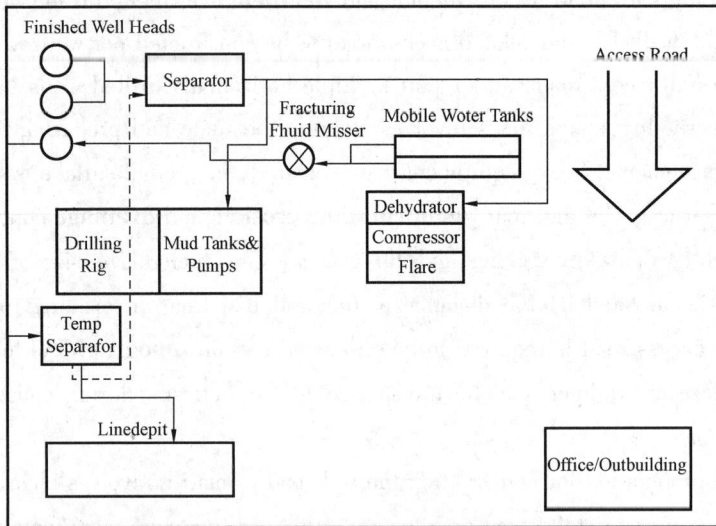

**Figure 4.3    Schematic of Typical Well Pad**

difference in horizontal drilling is the use of **downhole** motors to enable the drill bit to change direction.

### *Vertical Drilling*

Modern gas drilling, whether for horizontal or vertical wells, is typically performed using **rotary-drill rigs**, as shown in Figure 4.4. A hollow drilling line of **pipe**, or **drill string**, transmits torque from a drive to a weighted, rotating drill bit to penetrate the surface and underlying sediments. Drill bits are often **studded** with industrial diamonds to enable them to grind through any type rock.

Drilling fluids, or drilling mud, are pumped through the drill string to lubricate the drill bit, remove rock cuttings and waste from the wellbore, keep the wellbore open and stabilized, and to keep formation fluids from entering the wellbore during drilling. Drilling mud is a manufactured fluid composed of **synthetic oils**, polymer-based fluids, or water and **barite**. During drilling operations, well site engineers often blend chemical additives with drilling mud to achieve particular physical or chemical characteristics.

The mud carries the cuttings up the tubular space between the drill string and the sides of the wellbore to the surface where cuttings are filtered using a **shale shaker** or shale conveyor. Mud is then returned to the **mud pits** for treatment and eventual reuse. **Simultaneously**, the drill bit further recedes under-

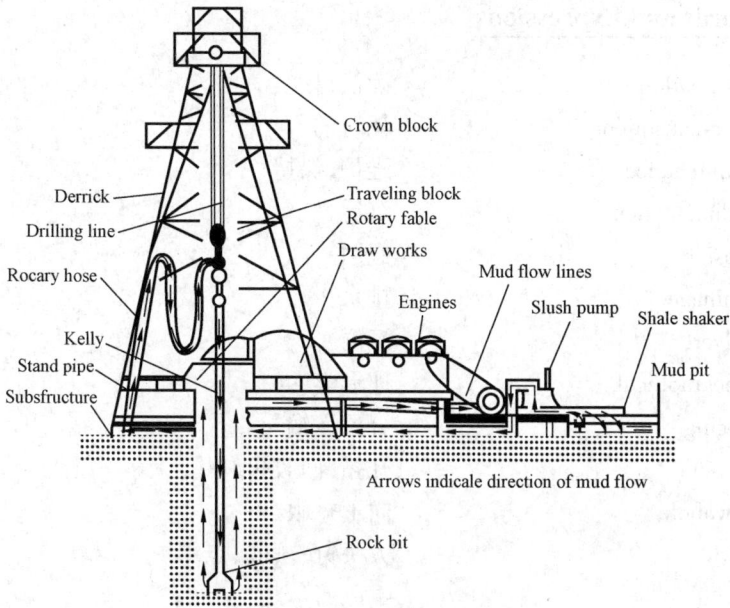

**Figure 4.4 Schematic of Rotary Drill**

ground and new sections of drill string are added while the bit continues to rotate until the well reaches planned depth. Meanwhile, the composition of drilling fluid is often altered to **impart** properties that minimize formation damage as the drill bit enters into the actual hydrocarbon reservoir.

*Horizontal Drilling*

**Directional drilling** enables wells to deviate from conventional vertical course. This is particularly beneficial in accessing gas shales as these formations typically demonstrate a more horizontal orientation. Horizontal drilling enables parallel intersection of the shale formation by the lower part, or horizontal lateral, of the wellbore.

Drilling a horizontal well increases contact with the reservoir; providing better access to the gas trapped inside and more efficient gas production. It additionally allows access to resources where vertical wells could be problematic, such as shale formations that run under heavily populated areas or geologic features such as a lake. The horizontal laterals in shale gas development may range from 1,000 to over 5,000 feet. Further, horizontal drilling makes drilling multiple wells from a single well pad possible; enabling E&P companies to reduce costs and surface disturbance associated with well pad construction.

## ⊞ Words and Expressions

| | |
|---|---|
| verification | 确认,核实 |
| time-consuming | 耗时的 |
| maintenance | 维护,保持 |
| sedimentation | 沉淀 |
| erosion | 侵蚀 |
| drainage | 排水 |
| culvert | 涵洞 |
| topographical | 地形学的 |
| staging | 分段运输 |
| reclaim | 开拓,回收再利用 |
| downhole | 向下打眼 |
| stud | 大头钉;镶嵌,点缀 |
| barite | 重晶石 |
| simultaneously | 同时地 |
| impart | 给予;告知 |

## ⊞ Phrases and Expressions

| | |
|---|---|
| development well | 开发井 |
| exploratory well | 探井 |
| drainage ditch | 排水沟 |
| gas gathering system | 集气系统 |
| exclusive of... | 不包括,除……之外 |
| rotary-drill rig | 旋转钻机 |
| drill pipe | 钻杆 |
| drill string | 钻柱 |
| synthetic oil | 合成油 |
| shale shaker | 泥浆振动筛 |
| mud pit | 泥浆池 |
| directional drilling | 定向钻井 |

## ⊞ Language Focus

1. Wells drilled vertically require one well pad and related access road to be

constructed per well, while one well pad and access road can support multiple horizontally drilled wells.

（参考译文：钻一口垂直井只需建造一个井场以及相关的道路即可，而一个井场及通道也可同时支持多个水平钻井。）

本句中，drilled vertically 是过去分词短语作 wells 的后置定语，while 是并列连词，意思是"而"，表示对比。

2. The surface area required for the well pad itself, exclusive of utility access and roads, ranges from two to five acres during drilling and fracturing to allow room for equipment movement and staging and containment pits.

（参考译文：不包括公用设施和道路占地在内，井场本身建设一般需要占地 2～5acre，以便钻井和压裂作业时有足够空间可以搬运设备，安放蓄水池等。）

本句为简单句，主谓结构是 The surface area ranges from two to five acres，主语后的 required for the well pad itself 是其后置定语，exclusive of utility access and road 也是主语的后置定语，to allow 不定式短语是目的状语。

3. As technological advances have improved their economics and operational feasibility, horizontal wells have become more favored by the industry for many shale plays as these wells enable increased pay zone contact.

（参考译文：随着技术的进步，直井和水平井的经济性及可行性都有所提高，水平井可以增加井和产层的接触面积，所以更受页岩气开发商的青睐。）

本句含有两个 as 引导的从句，As technological advances have improved their economics and operational feasibility 是时间状语从句；而 as these wells enable increased pay zone contact 为原因状语从句。

## 🔲 Reinforced Learning

### I. Answer the following questions for a comprehension of the text.

1. What is drilling mud?

2. What are the functions of drilling mud?

3. What does well site preparation involve?

4. Why have horizontal wells become more favored in shale gas extraction?

5. What are the differences between vertical and horizontal wells?

Ⅱ. **Multiple choice：choose the correct one from the alternative answers to give the exact meaning of the words.**

1. Formation evaluation, or well logging, is a commonly used technique in developed as well as exploratory wells.

A. estimation　　B. appropriation　　C. comment　　D. discussion

2. Exploratory wells are only drilled in areas where other data has indicated a high probability of accessible shale gas formations.

A. accelerant　　B. receptive　　C. acceptable　　D . approachable

3. Site preparation for constructing the well pad primarily involves clearing and leveling an appropriately sized area to support the drilling equipment footprint and related containment pits for drilling fluid and water and preparing the surface to support the movement of heavy equipment.

A. mainly　　B. privately　　C. vitally　　D. previously

4. The surface area required for the well pad itself, exclusive of utility access and roads, ranges from two to five acres during drilling and fracturing to allow room for equipment movement and staging and containment pits.

A. expected　　B. despite of　　C. beside　　D. except

5. Drilling fluids, or drilling mud, are pumped through the drill string to lubricate the drill bit, remove rock cuttings and waste from the wellbore, keep the wellbore open and stabilized, and to keep formation fluids from entering the wellbore during drilling.

A. steady　　B. moiled　　C. sturdy　　D. fixed

6. Simultaneously, the drill bit further recedes underground and new sections of drill string are added while the bit continues to rotate until the well reaches planned depth.

A. similarly　　　　　　　　B. meanwhile

C. synchronously　　　　　　D. likewise

7. This is particularly beneficial in accessing gas shales as these formations typically demonstrate a more horizontal orientation.

A. put　　B. show　　C. certify　　D. march

8. This is particularly beneficial in accessing gas shales as these formations typically demonstrate a more horizontal orientation.

A. intonation　　B. eastern　　C. direction　　D. location

9. It additionally allows access to resources where vertical wells could be

problematic, such as shale formations that run under heavily populated areas or geologic features such as a lake.

    A. diplomatic    B. debatable    C. confused    D. unknown

    10. Further, horizontal drilling makes drilling multiple wells from a single well pad possible; enabling E&P companies to reduce costs and surface disturbance associated with well pad construction.

    A. rubbish    B. confusion    C. influence    D. perturbation

**III. Multiple choice: read the four suggested translations and choose the best answer.**

    1. A number of exploratory wells may need to be analyzed in order to sufficiently characterize the potential of a shale formation, particularly if the geologic basin is large and there are variations in the target shale zone.

    A. 发掘    B. 描绘……的特征    C. 表示    D. 塑造

    2. Roads provide access to the site for well construction and maintenance.

    A. 坚持    B. 维修    C. 治理    D. 保护

    3. Simultaneously, the drill bit further recedes underground and new sections of drill string are added while the bit continues to rotate until the well reaches planned depth.

    A. 相反地    B. 同时地    C. 一样地    D. 渐渐地

    4. Drilling mud is a manufactured fluid composed of synthetic oils, polymer-based fluids, or water and barite.

    A. 生疏的    B. 半制成的    C. 已制成的    D. 成熟的

    5. Horizontal drilling enables parallel intersection of the shale formation by the lower part, or horizontal lateral, of the wellbore.

    A. 侧面    B. 正面    C. 斜面    D. 直线

**IV. Put the following sentences into Chinese.**

    1. The size of the access road is dictated by the distance of the well pad from an existing road, the route the access road is required to take to avoid certain topographical features, and the size of equipment to be transported to the well site during construction and drilling.

    2. Additionally, gas gathering systems are constructed to connect individual wells to pipelines for transporting gas.

    3. Similar technology and equipment are employed for drilling both vertical

and horizontal wells, however there are some differences.

4. During drilling operations, well site engineers often blend chemical additives with drilling mud to achieve particular physical or chemical characteristics.

5. Drilling a horizontal well increases contact with the reservoir; providing better access to the gas trapped inside and more efficient gas production.

**V. Put the following paragraphs into Chinese.**

1. The surface area required for the well pad itself, exclusive of utility access and roads, ranges from two to five acres during drilling and fracturing to allow room for equipment movement and staging and containment pits. During gas production, the area no longer required for these purposes is typically reclaimed, which reduces the area of the actual well pad to one to three acres.

2. The mud carries the cuttings up the tubular space between the drill string and the sides of the wellbore to the surface where cuttings are filtered using a shale shaker or shale conveyor. Mud is then returned to the mud pits for treatment and eventual reuse. Simultaneously, the drill bit further recedes underground and new sections of drill string are added while the bit continues to rotate until the well reaches planned depth. Meanwhile, the composition of drilling fluid is often altered to impart properties that minimize formation damage as the drill bit enters into the actualhydrocarbon reservoir.

# 4.3　Horizontal Drilling and Well Logging Methods

🔲 **Guidance to Reading**

*Horizontal drilling is to drill a well bore vertically to about the top of the producing formation, kick the well bore off vertical about 90°, and drill horizontally as far as possible in the producing formation. Compared to conventional steerable motors, **rotary steerable systems** (**RSS**) mitigate the technical challenges presented by stopping and starting of rotation and make it possible to navigate the drill bit while the bit rotates, offer more accurate **trajectory** control and reduce drilling time. Well logging involves making measurements in drill holes with probes designed to measure the physical and chemical properties of rocks and their contained fluids, while in mud logging much subsurface information can be obtained from samples of rock brought to the surface in cores or bit cuttings, or from other clues while drilling, such as penetration rate.*

## Text

### Conventional Steerable Motors

In horizontal drilling, changes in direction are accomplished using a steerable motor. The steerable motor assembly is attached to the drill string as part of the **bottom hole assembly** (**BHA**). The BHA is comprised of the drill bit, collars to apply weight to the drill bit, and stabilizers to keep the drilling assembly centered in the borehole. The steerable motor assembly consists of a power section to convert hydraulic power from the drilling mud into rotary motion, a bend section of up to 3 degrees to achieve non-vertical drill bit **orientation**, a **drive shaft**, and the drill bit itself.

Directional drilling using conventional steerable motors is accomplished in two modes: rotating ("rotating mode") and orienting (also referred to as "sliding mode"). In rotating mode, the drill string rotates to penetrate rock. To initiate a change in the wellbore direction, the rotation of the drill string is halted in such a position that the bend in the motor assembly points in the direction of the new trajectory and then **rotation** resumes.

Several issues leading to drilling and well completion inefficiencies are associated with the use of conventional steerable motors for horizontal drilling. When not rotating, the drill string can become stuck under the substantial weight of downhole cuttings and drilling mud. Alternation between rotating and sliding modes results in both a more tortuous path to the gas formation and rougher wellbore interior. These issues increase the time required to complete drilling and casing operations.

### Rotary Steerable Systems

Rotary steerable systems (RSS), commercialized in the mid – 1990s, combined with highly accurate measuring devices, have enabled directional drilling to become both more effective and economical and are key enablers of the rapid growth in shale gas production over the past decade. These systems enable the drill bit to continue rotating while changing angle or steering-an advancement over the stop and start operation of conventional steerable motors.

In RSS, a steering system is programmed with a planned wellpath developed by engineers and geologists. A steerable motor is attached to the drill string behind the drill bit. The BHA is equipped with sensing devices to continuously re-

cord and report survey data ( i. e. , depth, **inclination**, and compass direction),
which enable surface operators to track the exact location of the drill bit. **Pulses**
transmitted through the drilling mud enable the system to navigate and steer the
drill bit along the planned well path or initiate course corrections as needed.

RSS can be categorized into two types: "push-the-bit" and "point-the-bit"
systems. Push-the bit systems incorporate bias and control units into the BHA. The
**bias unit** applies force to "push" the bit in a controlled direction while the drill
string rotates. The control unit contains self-powered electronics, sensors, and a
mechanism to control the degree of force and direction of the bit side loads to
achieve the desired trajectory. Point-the-bit systems steer the bit by tilting it in the
direction of the desired angle.

Rotary steerable systems enable operators to control the direction of the drill
bit while the bit continuously rotates and offer greater trajectory control. Compared
to conventional steerable motors, RSS mitigate some of the technical challenges
presented by stopping and starting of rotation, reduce drilling time, and increase
the precision and accuracy of wellbore placement. The improvement in all of these
factors translates to economic benefits for operators.

Drilling efficiency, measured as total cost per meter, is a critical economic
factor in shale gas development. Continuous rotation while steering enables higher
**rates of penetration** ( **ROP** ) and provides a more steady steering model and
therefore greater control over the trajectory of the wellbore and smoother wellbores. By alleviating many of the challenges that typically occur with the use of
conventional steerable motors, RSS have enabled wellbores to accurately follow
long (2,500 to 9,000 feet) pay streaks and have improved drilling cost economics.

### Well Logging

Formation evaluation involves measuring and analyzing specific **petrophysi-
cal properties** of the rock in the immediate **vicinity** of a wellbore to determine a
formation's boundaries, volume of hydrocarbons, and ability to produce fluids to
the surface. Also known as well logging, the process enables a detailed record ( a
**well log**) of the geologic formations penetrated by a borehole and is conducted at
various phases of a well development including drilling, completion, production,
and abandonment. The well log may be based either on visual inspection of samples brought to the surface, geological logs, or on physical measurements made by

instruments lowered into the hole, geophysical logs.

### Mud Logging

Geological logs use data collected at the surface rather than by downhole instruments. The most commonly used geological logging technique in the oil and gas industry is mud logging. In this case logs are made by examining the bits of rock circulated to the surface by the drilling mud in rotary drilling. A critical parameter a typical **mud log** displays in shale developments is the amount of formation gas.

In addition to **lithology** (physical characteristics of rock samples), mud logs also typically include real-time drilling parameters such as rate of penetration, temperature of the drilling fluid, and chlorides. Data logged during the process may also include **mud weight**, pore pressure, directional data or **deviation surveys**, **weight on bit**, rotary speed, pump rate and pressure, viscosity, drill bit parameters, and **casing shoe** depth.

### Wireline Logging

In **geophysical logging**, electronic sensors are run through the wellbore to measure density and porosity, permeability, and resistivity. Imaging tools are run through the wellbore to record a picture of the formation along the well's length. **Acoustic logs** measure rock properties and help correlate data with previous seismic surveys. **Magnetic resonance** measurements characterize the volume and type of fluids in the formation as well as provide a direct measure of permeability. Nuclear logging instruments that actively bombard rocks with gamma rays are particularly useful in shale development as they enable accurate estimates of permeability and can be used in wells with **production casings**. **Nuclear magnetic resonance** (**NMR**) logging tools are an innovative new technology being used to measure rock **porosity**, estimate permeability from pore size distribution and identify pore fluids (water, oil and gas).

The two most common methods of geophysical data collection are wireline logging and **logging while drilling** (**LWD**). Wireline logging is conducted by pushing instruments through the wellbore after it is drilled. This technique can be used to evaluate well integrity, perform cement evaluations in the well casing process, and can be used in producing wells to perform production logging.

### Logging While Drilling

In the LWD process, instruments are attached to the drill string and take measurements while the well is being drilled. LWD measurements are taken in real-time before any damage has been sustained by the reservoir as a result of the drilling process and enable greater accuracy. Real-time measurements also enable "geo-steering" where geological markers identified by LWD tools are used to adjust the drill bit orientation and guide the wellbore to a specific target.

LWD tools that provide such directional information are also referred to as **measurement while drilling** (**MWD**) tools. In advanced systems, such MWD tools are built in to the rotary steerable system, allowing the tool to automatically alter its course based on a planned trajectory. The ability to measure, log and analyze down hole geophysical data in real time and ability to ensure that the well bore is **optimally** located within the hydrocarbon resource has been a key factor in the success of horizontal drilling in deep, low-permeability shale gas resources.

## Words and Expressions

| | |
|---|---|
| steerable | 可操纵的;易改变位置的 |
| trajectory | 轨道,轨线 |
| orientation | 定位,定向;方位 |
| rotation | 旋转,循环 |
| inclination | 倾向,斜坡 |
| pulses | 脉冲 |
| vicinity | 邻近,附近 |
| lithology | 岩石学,岩性 |
| porosity | 孔隙度 |
| optimally | 最佳,最适宜 |

## Phrases and Expressions

| | |
|---|---|
| rotary steerable systems (RSS) | 旋转转向系统 |
| bottom hole assembly (BHA) | 井下组装设备 |
| drive shaft | 驱动轴 |
| bias unit | 偏置器 |
| rates of penetration (ROP) | 钻进效率 |
| petrophysical property | 岩石物理性质 |
| well log | 钻井记录;录井记录;测井图; |

|  | 录井图;测井 |
| --- | --- |
| mud log | 泥浆录井 |
| mud weight | 泥浆比重 |
| deviational survey | 井斜测量 |
| weight on bit | 钻压 |
| casing shoe | 套管鞋 |
| geophysical logging | 地球物理测井 |
| acoustic log | 声波测井 |
| magnetic resonance | 磁共振 |
| production casing | 生产套管 |
| nuclear magnetic resonance（NMR） | 核磁共振 |
| logging while drilling（LWD） | 随钻测井 |
| measurement while drilling（MWD） | 随钻测量 |

## Language Focus

1. In advanced systems, such MWD tools are built in to the rotary steerable system, allowing the tool to automatically alter its course based on a planned trajectory.

（参考译文:在高级测井系统中,MDW 工具安装在旋转转向系统里,按照预设井眼轨迹自动调节钻进方向。）

本句中, allowing... 是现在分词短语做伴随状语, based on a planned trajectory 是过去分词短语做状语,修饰 alter its course。

2. Rotary steerable systems（RSS）, commercialized in the mid-1990s, combined with highly accurate **measuring** devices, have enabled directional drilling to become both more effective and economical and are key enablers of the rapid growth in shale gas production over the past decade.

［参考译文:旋转转向系统(RSS)是在 20 世纪90 年代中期实现商业化生产的,该系统配以高精度测量仪器,使定向钻井更加经济、有效,是过去十年里页岩气快速产增的关键因素。］

本句为简单句,主谓结构是 RSS have enabled... and are...,主语带两个谓语;commercialized in the mid-1990s 是过去分词短语做主语后置定语, combined with highly accurate measuring devices 是过去分词短语做状语,修饰谓语。本句有两个并列的谓语动词。

## Reinforced Learning

### I. Answer the following questions for a comprehension of the text.

1. What factors will lead to inefficiencies in the process of drilling and well completion?

2. For what reason are nuclear logging instruments useful in shale development?

3. How did RSS promote the rapid growth in shale gas production in the mid – 1990s?

4. Please describe the two types of RSS.

5. What are theparameters provided in the process of mud logging?

### II. Multiple choice: choose the correct one from the alternative answers to give the exact meaning of the words.

1. RSS can be <u>categorized</u> into two types: "push-the-bit" and "point-the-bit" systems.

    A. classified     B. amalgamated     C. combined     D. merged

2. Push-the bit systems <u>incorporate</u> bias and control units into the BHA.

    A. exclude     B. embody     C. remove     D. eliminate

3. Compass enables surface operators to <u>track</u> the exact location of the drill bit.

    A. trace     B. describe     C. draw     D. Seek

4. The most commonly used geological logging <u>technique</u> in the oil and gas industry is mud logging.

    A. technology          B. art

    C. handicraft          D. workmanship

5. Nuclear logging instruments that actively <u>bombard</u> rocks with gamma rays are particularly useful in shale development as they enable accurate estimates of permeability and can be used in wells with production casings.

    A. bomb     B. collapse     C. boost     D. boom

6. The ability to measure, log and analyze down hole geophysical data in real time and ability to <u>ensure</u> that the well bore is optimally located within the hydrocarbon resource has been a key factor in the success of horizontal drilling in deep, low-permeability shale gas resources.

A. make        B. favor        C. insist        D. guarantee

7. RSS <u>mitigate</u> some of the technical challenges presented by stopping and starting of rotation.

A. worsen        B. relieve        C. deter        D. aggravate

8. By <u>alleviating</u> many of the challenges that typically occur with the use of conventional steerable motors, RSS have enabled wellbores to accurately follow long (2,500 to 9,000 feet) pay streaks and have improved drilling cost economics.

A. relieving                   B. adding

C. transmitting             D. transforming

9. Pulses transmitted through the drilling mud enable the system to navigate and steer the drill bit along the planned well path or <u>initiate</u> course corrections as needed.

A . make        B. start        C. end        D. plan

10. LWD tools are used to adjust the drill bit orientation and guide the wellbore to a <u>specific</u> target.

A. particular        B. general        C. broad        D . small

**III. Multiple choice: read the four suggested translations and choose the best answer.**

1. Several issues leading to drilling and well completion <u>inefficiencies</u> are associated with the use of conventional steerable motors for horizontal drilling.

A. 毫无效率        B. 高效        C. 效率        D. 作用

2. When not rotating, the drill string can become stuck under the <u>substantial</u> weight of downhole cuttings and drilling mud.

A. 大量的,重大的           B. 微小的

C. 结实的,牢固的           D. 地下的

3. Rotary steerable systems (RSS), <u>commercialized</u> in the mid – 1990s, combined with highly accurate measuring devices, have enabled directional drilling to become both more effective and economical and are key enablers of the rapid growth in shale gas production over the past decade.

A. 福利化        B. 商业化        C. 系统化        D. 精细化

4. Pulses transmitted through the drilling mud enable the system to <u>navigate</u> and steer the drill bit along the planned well path or initiate course corrections as needed.

A. 操纵　　　　　B. 控制　　　　　C. 指挥　　　　　D. 导航

5. Formation evaluation involves measuring and analyzing specific petro-physical <u>properties</u> of the rock in the immediate vicinity of a wellbore to determine a formation's boundaries, volume of hydrocarbons, and ability to produce fluids to the surface.

A. 财富　　　　　B. 质量　　　　　C. 属性　　　　　D. 体积

6. In addition to lithology (physical characteristics of rock samples), mud logs also typically include <u>real-time</u> drilling parameters.

A. 确实的　　　　B. 实时的　　　　C. 多样的　　　　D. 单一的

7. In advanced systems, such MWD tools are built in to the rotary steerable system, allowing the tool to <u>automatically</u> alter its course based on a planned trajectory.

A. 机械的　　　　B. 自动地　　　　C. 被动地　　　　D. 主动地

8. The well log may be based either on <u>visual</u> inspection of samples brought to the surface, geological logs, or on physical measurements made by instruments lowered into the hole, geophysical logs.

A. 视觉的　　　　B. 听觉的　　　　C. 触觉的　　　　D. 不同的

9. This technique can be used to <u>evaluate</u> well integrity, perform cement evaluations in the well casing process, and can be used in producing wells to perform production logging.

A. 猜测　　　　　B. 评测　　　　　C. 检查　　　　　D. 勘察

10. These systems enable the drill bit to continue rotating while changing <u>angle</u> or steering-an advancement over the stop and start operation of conventional steerable motors.

A. 角度　　　　　B. 天使　　　　　C. 方向　　　　　D. 地位

**IV. Put the following sentences into Chinese.**

1. Alternation between rotating and sliding modes results in both a more tortuous path to the gas formation and rougher wellbore interior.

2. Compared to conventional steerable motors, RSS mitigate some of the technical challenges presented by stopping and starting of rotation, reduce drilling time, and increase the precision and accuracy of wellbore placement.

3. To initiate a change in the wellbore direction, the rotation of the drill string is halted in such a position that the bend in the motor assembly points in the direction of the new trajectory and then rotation resumes.

4. Several issues leading to drilling and well completion inefficiencies are associated with the use of conventional steerable motors for horizontal drilling.

5. Magnetic resonance measurements characterize the volume and type of fluids in the formation as well as provide a direct measure of permeability.

### V. Put the following paragraphs into Chinese.

1. Geological logs use data collected at the surface rather than by downhole instruments. The most commonly used geological logging technique in the oil and gas industry is mud logging. In this case logs are made by examining the bits of rock circulated to the surface by the drilling mud in rotary drilling. A critical parameter a typical mud log displays in shale developments is the amount of formation gas.

2. The two most common methods of geophysical data collection are wireline logging and logging while drilling (LWD). Wireline logging is conducted by pushing instruments through the wellbore after it is drilled. This technique can be used to evaluate well integrity, perform cement evaluations in the well casing process, and can be used in producing wells to perform production logging.

## 4. 4  Well Completion

### Guidance to Reading

*In oil and gas production, completion is the process of making a well ready for production. This principally involves preparing the bottom of the hole to the required **specifications**, running in the **production casing** and **tubing** and the associated down hole tools, as well as perforating and stimulating as required. **Cased-hole** perforated **completions** are commonly used in shale gas development. However, advances in completion technology have led to the emergence of **open-hole** multi-stage fracturing systems for horizontal wells and open-hole completions have also been utilized in many shale plays. While stimulation treatments vary depending on formation type, hydraulic fracturing is the type of stimulation treatment used to open up the shale formation, enabling the efficient flow of gas from the rock.*

## ⬚ Text

### *Well Casing*

Casing creates a barrier to restrict fluid transfer between the well and surrounding rock; confining gas from leaching out of and water from flooding into the well. This provides several benefits. Chief among them is ensuring that the gas accessed from the reservoir can be recovered for sale and protecting surrounding formations such as fresh water zones from contamination. Casing also provides stability to the upper formations of the wellbore to prevent it from caving, seals the surface from high pressure zones to prevent blowout, and creates a smooth internal bore for installing production tubing and equipment.

Casing a well involves the installation of **concentric** layers of large diameter steel pipe and cement within the borehole. After each layer of casing pipe is set, it is cemented to ensure a seal between the casing and the formation and between the two pipes. Each layer of pipe performs a specific purpose. The typical gas well casing is:

- **Conductor casing** (outermost pipe) prevents collapse of loose soil near the surface and enables circulation of drilling fluids during drilling. Usually 16 to 20 inches in **diameter**, about 50 feet long, and installed prior to the start of drilling operations.

- **Surface casing** protects **freshwater zones** near the surface from **contamination** from leaching fluids during drilling and gas during production; the length of casing **is dictated by** the depth of the **freshwater aquifer**.

- **Intermediate casing** is used in certain sections to protect the well from various **hazardous** subsurface conditions such as abnormal pressure zones and salt water deposits.

- Production casing (innermost pipe) provides a **conduit** for the production tubing, which is ultimately inserted inside the production casing to carry the extracted gas to the surface.

To prevent fluid transfer between aquifers and the wellbore, steel surface casings are inserted and cemented into the well to depths of between 1,000 and 4,000 feet. Once the drill bit has passed the **water table**, the drill string is tripped out of the wellbore and surface casing is strung in. The **annular space** between the surface casing string and drilled hole is filled with cement. Once the cement has set, drilling of the wellbore is continued from the bottom of the first cemented steel casing to the next depth. This process is repeated using smaller steel intermediate and production casings each time until the gas shale is reached.

After casing is cemented in, a cement evaluation is conducted. An electronic probe is used to measure the thickness of the cement to **validate** it will function as designed. In the United States, casing and cement specifications and process, or casing and cementing programs, are governed by state and federal regulations. Additionally, the production casing is tested to ensure the casing will hold pressure and contain the fracturing fluids used in hydraulic fracturing. Once these tests are successfully accomplished, the next steps in preparing the well for production are well completion and stimulation.

### Well Completion and Stimulation

Completion and stimulation activities enable the flow of natural gas out of the formation and up to the surface. Completion essentially involves preparing the bottom of the well hole to specifications designed to maximize gas production, installing production tubing and related downhole tools, perforating and stimulating the well (collectively called "**lower completion**") and installation of wellhead equipment. Stimulation is a treatment performed on hydrocarbon-bearing formations to improve the flow of gas from the formation to the wellbore. While stimulation treatments vary depending on formation type, hydraulic fracturing is the type stimulation treatment used to open up the shale formation, enabling the efficient flow of gas from the rock.

### Completion Activities

There are several types of completion techniques used in oil and gas development. Selection of the completion technique is largely dictated by the technical and economic viability of the approach to effectively intake gas given the location and characteristics of the hydrocarbon formation that the well is accessing. Generally, one of the two techniques is used in shale gas well completions:

● Cased-hole, perforated completion, where production casing and cement extend to the toe (bottom tip) of the well, the walls of which are perforated in the production zone using a shaped **explosive charge prior to** hydraulic fracturing

● Open-hole completion, in which no casing is cemented into place within the reservoir, the end of the piping is left open and gas is collected through the **production liner**.

Cased-hole perforated completions are commonly used in shale gas development. However, advances in completion technology have led to the emergence of open-hole multi-stage fracturing systems (OHMS) for horizontal wells and open-hole completions have also been utilized in many US shale plays. Open-hole completions enable the elimination of costs associated with casing, cementing, and perforating. Further, they have been found to enable higher initial gas production rates from shale reservoirs than conventional perforated completions. However, some degree of control over the placement of fracture intervals and the fracture path is lost compared to cased-hole completions.

Before fracturing operations, a wellhead, or "frac tree" designed specifically for hydraulic fracturing is installed. Additionally, flowback equipment including pipes, **manifolds**, and a gas-water separator, are installed at the surface and the system is pressure tested. After fracturing, the frac tree is replaced with a high-pressure wellhead that remains in place to monitor and regulate the flow of gas and other fluids out of the well during gas production.

### Hydraulic Fracturing

Due to their low porosity and low permeability, shales do not release their gas freely. Therefore, shales are hydraulically fractured during well completion operations to enable efficient gas recovery. Hydraulic fracturing, also referred to variously as "hydrofracking" and "fracking," involves isolating sections of the well in the producing zone and then pumping high volumes of specially engineered fracturing fluids at high pressure (up to 8,000 psi) down the wellbore and out into the shale formation. Internal fluid pressure opens the fractures and causes them to **propagate** through the rock.

The creation of fractures simultaneously increases the surface area of a formation exposed to the borehole and establishes a conductive path connecting the reservoir and the well. These paths increase the rate that fluids can be produced from the reservoir formations, in some cases by many hundreds of percent. Proppants, such as sand with a specific grain size, mixed in the fracturing fluid settle in the fractures and essentially prop the fractures open to allow continual gas flow. To maximize reservoir contact, the entire length of the horizontal portion of the wellbore must be optimally stimulated. Shales which are already naturally fractured receive hydraulic fracturing treatment in order to join the natural fractures to one another and create paths to the wellbore.

Cased-hole completions generate expenses that can prove uneconomic in some shale formations. Fracturing each stage in this way requires multiple **perforating guns** and **coiled tubing** trips and, therefore, repeated rigging up and down of stimulation equipment during operations. Open-hole multistage (OHMS) completions have become more technically viable in recent years. This technique reduces or eliminates costs associated with cementing, perforation, and isolation and reduces personnel and equipment costs by reducing the time required to complete fracturing operations.

## Words and Expressions

| | |
|---|---|
| specification | 规格,详述,说明书 |
| tubing | 油管 |
| concentric | 同心的 |
| diameter | 直径 |
| hazardous | 冒险的,有危险的,碰运气的 |
| contamination | 污染,弄脏,毒害,玷污 |
| conduit | 导管,水道,沟渠 |
| validate | 使⋯⋯有效;确认 |
| manifold | 管汇 |
| propagate | 繁衍;使遗传;扩散 |

## Phrases and Expressions

| | |
|---|---|
| production casing | 生产套管 |
| cased-hole completion | 套管完井 |
| open-hole | 裸眼井 |
| conductor casing | 导管 |
| surface casing | 表层套管 |
| freshwater aquifer/zone | 淡水层 |
| be dictated by | 由⋯⋯决定 |
| intermediate casing | 中间套管 |
| water table | 潜水面 |
| annular space | 环形空间 |

| lower completion | 底部/下部完井 |
|---|---|
| explosive charge | 炸药;射孔弹 |
| prior to | 在……之前 |
| production liner | 生产尾管 |
| perforating gun | 射孔抢 |
| coiled tubing | 挠性管 |

## Language Focus

1. Production casing (innermost pipe) provides a conduit for the production tubing, which is ultimately inserted inside the production casing to carry the extracted gas to the surface.

[参考译文:生产套管(最里层套管)为油管导管,油管最后下入到生产套管内,将采出气输送到地面。]

本句中 which 引导非限制性定语从句,修饰 the production tubing。

2. Completion essentially involves preparing the bottom of the well hole to specifications designed to maximize gas production, installing production tubing and related down hole tools, perforating and stimulating the well (collectively called "lower completion") and installation of wellhead equipment.

[参考译文:完井基本上是按设计规格进行井底作业以实现产气量最大化,安装油管和相关井下设备,射孔并增产(总称"低部完井"),同时安装井口设备。]

本句中,谓语动词 involve 后面有四个并列的宾语,分别是 preparing、installing、perforating and stimulating 和 installation,前面三个都是动宾结构构成的动名词短语;preparing 动名词短语中的 designed to 过去分词短语是后置定语,修饰 specifications。

## Reinforced Learning

I. Answer the following questions for a comprehension of the text.

1. What benefits does well casing provide?

2. Whatdoes the typical gas well casing consist of ?

3. Please introduce two techniquesused in shale gas well completion.

4. What are the advantages of Open-hole completions in shale play?

5. What is the purpose of well completion and stimulation?

**II. Multiple choice:choose the correct one from the alternative answers to give the exact meaning of the words.**

1. Surface casing protects freshwater zones near the surface from <u>contamination</u> from leaching fluids during drilling and gas during production.

A. contemporary    B. combination    C. pollution    D. violation

2. The length of casing is <u>dictated</u> by the depth of the freshwater aquifer.

A. dated    B. decided    C. districted    D. approved

3. Selection of the completion technique is largely dictated by the technical and economic <u>viability</u> of the approach to effectively intake gas given the location and characteristics of the hydrocarbon formation that the well is accessing.

A. feasibility    B. capacity

C. compability    D. accommodation

4. Casing creates a <u>barrier</u> to restrict fluid transfer between the well and surrounding rock.

A. obstacle    B. door    C. method    D. emigrants

5. After casing is cemented in,a cement <u>evaluation</u> is conducted.

A. assessment    B. appreciation    C. judge    D. test

6. After fracturing,the frac tree is replaced with a high-pressure wellhead that remains in place to <u>monitor</u> and regulate the flow of gas and other fluids out of the well during gas production.

A. distinguish    B. supervise    C. commence    D. associate

7. Before fracturing operations,a wellhead,or "frac tree" designed specifically for hydraulic fracturing is <u>installed</u>.

A. extended    B. expanded    C. fixed    D. shortened

8. To prevent fluid transfer between aquifers and the wellbore,steel surface casings are <u>inserted</u> and cemented into the well to depths of between 1,000 and 4,000 feet.

A. component    B. constructed    C. instituted    D. embedded

9. An electronic probe is used to measure the thickness of the cement to <u>validate</u> it will function as designed.

A. prove    B. reform    C. transform    D. refer

10. In the United States,casing and cement specifications and process,or casing and cementing programs,are <u>governed</u> by state and federal regulations.

A. controled    B. measured    C. increased    D. decreased

**III. Multiple choice：read the four suggested translations and choose the best answer.**

1. Additionally, the production casing is tested to ensure the casing will hold pressure and contain the <u>fracturing fluids</u> used in hydraulic fracturing.

　A. 压裂液　　　　B. 稀释液　　　　C. 酸液　　　　D. 返排液

2. The <u>annular</u> space between the surface casing string and drilled hole is filled with cement.

　A. 环形的　　　　B. 矩形的　　　　C. 三角形的　　D. 多边形的

3. Selection of the completion technique is largely dictated by the technical and economic viability of the approach to effectively <u>intake</u> gas given the location and characteristics of the hydrocarbon formation that the well is accessing.

　A. 吐出　　　　B. 吸入　　　　C. 镶嵌　　　　D. 拥抱

4. An electronic probe is used to measures the thickness of the cement to <u>validate</u> it will function as designed.

　A. 起作用　　　　B. 运转　　　　C. 工作　　　　D. 行使职责

5. Fracturing each stage in this way requires <u>multiple</u> perforating guns and coiled tubing trips and, therefore, repeated rigging up and down of stimulation equipment during operations.

　A. 有限的　　　　B. 减少的　　　　C. 多重的　　　D. 无限的

**IV. Put the following sentences into Chinese.**

1. An electronic probe is used to measure the thickness of the cement to validate it will function as designed.

2. Completion and stimulation activities enable the flow of natural gas out of the formation and up to the surface.

3. Due to their low porosity and low permeability, shales do not release their gas freely.

4. After fracturing, the frac tree is replaced with a high-pressure wellhead that remains in place to monitor and regulate the flow of gas and other fluids out of the well during gas production.

5. While stimulation treatments vary depending on formation type, hydraulic fracturing is the type of stimulation treatment used to open up the shale formation, enabling the efficient flow of gas from the rock.

**V. Put the following paragraphs into Chinese.**

1. Casing creates a barrier to restrict fluid transfer between the well and surrounding rock; confining gas from leaching out of and water from flooding into the well. This provides several benefits. Chief among them is ensuring that the gas accessed from the reservoir can be recovered for sale and protecting surrounding formations such as fresh water zones from contamination. Casing also provides stability to the upper formations of the wellbore to prevent it from caving, seals the surface from high pressure zones to prevent blowout, and creates a smooth internal bore for installing production tubing and equipment.

2. The creation of fractures simultaneously increases the surface area of a formation exposed to the borehole and establishes a conductive path connecting the reservoir and the well. These paths increase the rate that fluids can be produced from the reservoir formations, in some cases by many hundreds of percent. Prop pants, such as sand with a specific grain size, mixed in the fracturing fluid settle in the fractures and essentially prop the fractures open to allow continual gas flow. To maximize reservoir contact, the entire length of the horizontal portion of the wellbore must be optimally stimulated. Shales which are already naturally fractured receive hydraulic fracturing treatment in order to join the natural fractures to one another and create paths to the wellbore.

# Chapter 5  Production Impact
# on the Environment

## 5.1  Water Use in Fracturing

### ⌗ Guidance to Reading

*The current practice for hydraulic fracturing treatments of shale gas reservoirs is to apply a sequenced pumping event in which millions of gallons of water-based fracturing fluids mixed with proppant materials and thickening agents are pumped in a controlled and monitored manner into the target shale formation. The fracturing fluids used for gas shale stimulations consist primarily of water but also include a variety of chemical additives, which may result in contamination. The number of additives used varies depending on the conditions of the specific well being fractured. In **water-stressed areas**, the extraction of water for drilling and hydraulic fracturing can have abroad and serious environmental problems, such as lowering the water table, affecting **biodiversity** and harming the local **ecosystem**.*

### ⌗ Text

The extent of water use and the risk of watercontamination are key issues for any unconventional gas development and have generated considerable public concern. In the case of a shale gas or tight gas development, though some water is required during the drilling phase, the largest volumes of water are used during the hydraulic fracturing process: each well might need anything between a few thousand and 20,000 cubic metres (between 1 million and 5 million gallons). Efficient use of water during fracturing is essential. Average water use per well completion in the Eagle Ford play in west Texas has been reduced from 18.5 to 13.6 thousand cubic metres since mid −2010, primarily through increased recycling of waste water from flow-back of fracturing fluid, an important step forward, given that more than 2,800 drilling **permits** were issued by the Railroad Commission of Texas for Eagle Ford wells in 2011. The amount of water required for shale gas or tight gas developments, calculated per unit of energy produced, is higher than for conventional gas but comparable to the amount used for the production of conventional oil.

Water for fracturing can come from surface water sources, or from local boreholes (which may draw from shallow or deep aquifers and which may already have been drilled to support production operations), or from further **afield** (which generally requires trucking). Transportation of water from its source and to disposal locations can be a large-scale activity. If the hydraulic fracturing of a well requires 15000 cubic metres, this amounts to 500 truck-loads of water, on the basis that a typical truck can hold around 30 cubic metres of water. Such transportation **congests** local roads, increases **wear and tear** to roads and bridges and, if not managed safely, can increase road accidents.

In areas of **water-scarcity**, the extraction of water for drilling and hydraulic fracturing can have broad and serious environmental effects. It can lower the water table, affect biodiversity and harm the local ecosystem. It can also reduce the **availability** of water for use by local communities and in other productive activities, such as agriculture.

Limited availability of water for hydraulic fracturing could become a significant **constraint** on the development of tight gas and shale gas in some water-stressed areas. In China, for example, the Tarim Basin in the Xinjiang Uyghur Autonomous Region holds some of the country's largest shale gas deposits, but also suffers from severe water scarcity. Although not on the same scale, **in terms of** either resource endowment or water stress, a number of other prospective deposits occur in regions that are already experiencing intense competition for water resources. The development of China's shale gas industry has to date focused on the Sichuan basin, in part because water is much more abundant in this region.

Hydraulic fracturing **dominates** the freshwater requirements for unconventional gas wells and the dominant choice of fracturing fluid for shale gas, "**slickwater**", which is often available at the lowest cost and in some shale reservoirs may also bring some gas-production benefits, is actually the most **demanding** in terms of water needs. Much attention has accordingly been given to approaches which might reduce the amount of water used in fracturing. Total pumped volumes (and therefore water volumes required) can be decreased through the use of more traditional, high viscosity fracturing fluids (using **polymers** or **surfactants**), but these require a complex **cocktail** of chemicals to be added.

Foamed fluids, in which water is foamed with **nitrogen** or $CO_2$, with the help of surfactants (as used in dish washing liquids), can be attractive, as 90% of the fluid can be gas and this fluid has very good proppant-carrying properties. Water can, indeed, be eliminated altogether by using **hydrocarbon-based fracturing fluids**, such as propane or **gelled hydrocarbons**, but their flammability makes them more difficult to handle safely at the well site. The percentage of fracturing fluid that gets back-produced during the flow-back phase varies with the type of fluid used (and the shale characteristics), so the optimum choice of fluid will depend on many factors: the availability of water, whether water recycling is included in the project, the properties of the shale reservoir being tapped, the desire to reduce the usage of chemicals and the economics.

### *What is in a fracturing fluid*?

Environmental concerns have focused on the fluid used for hydraulic fracturing and the risk of water contamination through leaks of this fluid into groundwater. Water itself, together with sand or ceramic beads (the "proppant"), makes up over 99% of a typical fracturing fluid, but a mixture of chemical additives is also used to give the fluid the properties that are needed for fracturing. These properties vary according to the type of formation. Additives (not all of which would be used in all fracturing fluids) typically help to accomplish four tasks:

• To keep the proppant suspended in the fluid while it is being pumped into the well and to ensure that the proppant ends up in the fractures being created. Without this effect, the heavier proppant particles would tend to be distributed unevenly in the fluid under the influence of gravity and would, therefore, be less effective. **Gelling** polymers, such as guar or cellulose (similar to those used in food and cosmetics) are used at a concentration of about 1%. **Cross-linking agents**, such as borates or metallic salts, are also commonly used at very low concentration to form a stronger gel. They can be toxic at high concentrations, though they are often found at low natural concentrations in mineral water.

• To change the properties of the fluid over time. Characteristics that are needed to deliver the proppant deep into subsurface cracks are not desirable at other stages in the process, so there are additives that give time-dependent properties to the fluid, for example, to make the fluid less viscous after fracturing, so that the hydrocarbons flow more easily along the fractures to the well. Typically, small concentrations of **chelants**（such as those used to **de-scale** kettles）are used, as are small concentrations of oxidants or enzymes（used in a range of industrial processes）to break down the gelling polymer at the end of the process.

● To reduce friction and therefore reduce the power required to inject the fluid into the well. A typical **drag-reducing polymer** is **polyacrylamide**（widely used, for example, as an absorbent in baby diapers）.

● To reduce the risk that naturally occurring **bacteria** in the water affect the performance of the fracturing fluid or **proliferate** in the reservoir, producing **hydrogen sulphide**; this is often achieved by using a **disinfectant**（**biocide**）, similar to those commonly used in hospitals or cleaning supplies.

## Words and Expressions

| | |
|---|---|
| biodiversity | 生物多样性 |
| ecosystem | 生态系统 |
| permit | 许可证 |
| afield | 远离着（家乡）;在（或去）远处;在野外 |
| congests | 使充满;使拥塞;使充血乎 |
| availability | 可得性,有效;可用度 |
| constraint | 约束;强制;约束条件 |
| dominate | 统治;主宰;控制 |
| demanding | 费力的;要求高的 |
| polymers | 聚合物 |
| surfactants | 表面活性剂 |
| cocktail | 混合物 |
| nitrogen | 氮 |
| gelling | 胶凝 |
| gel | 凝胶 |
| chelant | 螯合剂 |
| de-scale | 除垢 |
| polyacrylamide | 聚丙烯酰胺 |
| bacteria | 细菌 |
| proliferate | 增殖（增生,使……激增,使……扩散,迅速扩大） |

| | |
|---|---|
| disinfectant | 消毒剂 |
| biocide | 抗微生物剂　生物杀伤剂 |

## Phrases and Expressions

| | |
|---|---|
| water-stressed area | 缺水地区 |
| wear and tear | 磨损 |
| water-scarcity | 缺水 |
| in terms of | 就……而言 |
| slick-water | 滑溜水 |
| hydrocarbon-based fracturing fluid | 烃基压裂液 |
| gelled hydrocarbon | 胶态烃;稠化烃 |
| cross-linking agent | 交联剂 |
| drag-reducing polymer | 减租聚合物 |
| hydrogen sulphide | 硫化氢 |

## Language Focus

1. The amount of water required for shale gas or tight gas developments, calculated per unit of energy produced, is higher than for conventional gas but comparable to the amount used for the production of conventional oil.

（参考译文:开发页岩气或致密气藏所需的水量,若以生产单位能量的水耗来计算,开发页岩气或致密气藏所需的水量比常规气藏多,但同常规油藏相近。）

本句为复合句,带有一个 than 引导的比较状语从句,主句是主、系、表结构:The amount of water is higher than... but comparable to,表语部分为 but 连接的两个并列结构,but 后省略了 is,than 后省略了 the amount of water is required。主语后的 required for shale gas or tight gas developments 和 calculated per unit of energy produced 都是过去分词短语做其后置定语。第二个 the amount 后的 used for the production of conventional oil 是过去分词短语做其后置定语。

2. Hydraulic fracturing dominates the freshwater requirements for unconventional gas wells and the dominant choice of fracturing fluid for shale gas, "slickwater", which is often available at the lowest cost and in some shale reservoirs may also bring some gas-production benefits, is actually the most demanding in terms of water needs.

（参考译文：非常规气井对淡水的需求取决于水力压裂，页岩气开发所采用的主要压裂液"滑溜水"通常成本最低，在某些页岩气储集层中也能起到增产的作用，但其耗水量最高。）

本句为 and 连接的并列句，前面分句的主谓结构是 Hydraulic fracturing dominates the freshwater requirements，后面分句的主谓结构是 the dominant choice of. . . is actually the most demanding，主语后的"slick-water"是其同位语，which 引导非限制性定语从句，which 指代"slick-water"，在从句中做主语，后面带两个并列谓语，is often available 和 may also bring some gas-production benefits。

3. The percentage of fracturing fluid that gets back-produced during the flow-back phase varies with the type of fluid used（and the shale characteristics）, so the optimum choice of fluid will depend on many factors：the availability of water, whether water recycling is included in the project, the properties of the shale reservoir being tapped, the desire to reduce the usage of chemicals and the economics.

［参考译文：返排压裂液的比例与使用的压裂液种类（以及页岩性质）有关，因此压裂液的最佳选择需要考虑诸多因素：水是否容易获得、水是否循环利用、页岩气藏的特性、化学制剂的使用是否要减少以及其他经济因素。］

本句为复合句，主句主谓结构为 The percentage of fracturing fluid. . . varies with. . . , 其中定语从句 that gets back-produced during the flow-back phase 修饰其先行词 The percentage of fracturing fluid。在 so 引导的结果状语从句中，主谓结构是 the optimum choice of fluid will depend on many factors，后面列举了四个因素。

## Reinforced Learning

### I. Answer the following questions for a comprehension of the text.

1. Which requires higher amount of water in terms of per unit of energy, unconventional gas developments or conventional gas developments?

2. For what reason has the development of China's shale gas industry to date focused onthe Sichuan basin according to the text?

3. What dominates the freshwater requirements for unconventional gas wells?

4. What are foamed fluids and the advantages?

5. What is the composition of fracturing fluid?

Ⅱ. **Multiple choice：choose the correct one from the alternative answers to give the exact meaning of the words.**

1. In areas of water-scarcity, the underline{extraction} of water for drilling and hydraulic fracturing can have broad and serious environmental effects.

A. use       B. taking out       C. disposal       D. recycling

2. It can also reduce the underline{availability} of water for use by local communities and in other productive activities, such as agriculture.

A. openness       B. convenience       C. accessibility       D. approval

3. The development of China's shale gas industry has to date focused on the Sichuan basin, in part because water is much more underline{abundant} in this region.

A. unique       B. rare       C. rich       D. common

4. Total pumped volumes (and therefore water volumes required) can be underline{decreased} through the use of more traditional, high viscosity, fracturing fluids (using polymers or surfactants), but these require a complex cocktail of chemicals to be added.

A. reduced       B. promoted       C. progressed       D. defused

5. Foamed fluids, in which water is foamed with nitrogen or $CO_2$, with the help of surfactants (as used in dish washing liquids), can be attractive, as 90% of the fluid can be gas and this fluid has very good proppant-carrying underline{properties}.

A. fortunes       B. problems       C. minutes       D. attributes

6. Water can, indeed, be underline{eliminated} altogether by using hydrocarbon-based fracturing fluids, such as propane or gelled hydrocarbons, but their flammability makes them more difficult to handle safely at the well site.

A. experienced       B. concentrated       C. explored       D. removed

7. The underline{optimum} choice of fluid will depend on many factors.

A. best       B. minor       C. big       D. fundamental

8. Water itself, together with sand or ceramic beads (the "proppant"), underline{makes up} over 99% of a typical fracturing fluid, but a mixture of chemical additives is also used to give the fluid the properties that are needed for fracturing.

A. comprises       B. consists of       C. focuses       D. selects

9. Additives (not all of which would be used in all fracturing fluids) typically help to underline{accomplish} four tasks.

A. comply       B. complete       C. transform       D. refer

10. They can be underline{toxic} at high concentrations, though they are often found at

low natural concentrations in mineral water.

    A. optional    B. removable    C. delicious    D. poisonous

**Ⅲ. Multiple choice:read the four suggested translations and choose the best answer.**

1. In China,for example,the Tarim Basin in the Xinjiang Uyghur Autonomous Region holds some of the country's largest shale gas deposits,but also suffers from severe water scarcity.

    A. 大陆的    B. 自治的    C. 特权的    D. 地区的

2. Much attention has accordingly been given to approaches which might reduce the amount of water used in fracturing.

    A. 未来    B. 精华    C. 途径    D. 现实

3. To reduce friction and therefore reduce the power required to inject the fluid into the well.

    A. 碰撞    B. 报答    C. 接触    D. 摩擦

4. Water can,indeed,be eliminated altogether by using hydrocarbon-based fracturing fluids,such as propane or gelled hydrocarbons,but their flammability makes them more difficult to handle safely at the well site.

    A. 易燃性    B. 易腐蚀性    C. 耐高温性    D. 耐寒性

5. Environmental concerns have focused on the fluid used for hydraulic fracturing and the risk of water contamination through leaks of this fluid into groundwater.

    A. 预见    B. 关注    C. 事件    D. 建设

**Ⅳ. Put the following sentences into Chinese.**

1. The extent of water use and the risk of water contamination are key issues for any unconventional gas development and have generated considerable public concern.

2. They can be toxic at high concentrations,though they are often found at low natural concentrations in mineral water.

3. Limited availability of water for hydraulic fracturing could become a significant constraint on the development of tight gas and shale gas in some water-stressed areas.

4. The percentage of fracturing fluid that gets back-produced during the flow-back phase varies with the type of fluid used(and the shale characteristics),so the

optimum choice of fluid will depend on many factors: the availability of water, whether water recycling is included in the project, the properties of the shale reservoir being tapped, the desire to reduce the usage of chemicals and the economics.

5. Environmental concerns have focused on the fluid used for hydraulic fracturing and the risk of water contamination through leaks of this fluid into groundwater.

**V. Put the following paragraphs into Chinese.**

1. In areas of water-scarcity, the extraction of water for drilling and hydraulic fracturing can have broad and serious environmental effects. It can lower the water table, affect biodiversity and harm the local ecosystem. It can also reduce the availability of water for use by local communities and in other productive activities, such as agriculture.

2. Water for fracturing can come from surface water sources, or from local boreholes, or from further afield(which generally requires trucking). Transportation of water from its source and to disposal locations can be a large-scale activity. If the hydraulic fracturing of a well requires 15,000 cubic metres, this amounts to 500 truck-loads of water, on the basis that a typical truck can hold around 30 cubic metres of water. Such transportation congests local roads, increases wear and tear to roads and bridges and, if not managed safely, can increase road accidents.

# 5.2　Fracturing Fluid Additives

## 凸 Guidance to Reading

*Because the make-up of each fracturing fluid varies to meet the specific needs of each area, there is no one-size-fits-all formula for the volumes for each additive. Fracturing fluids are generally classified into three types: aqueous-based, oil, and **foam fluids**. Aqueous-based ones have been widely used because of their low cost, high performance, greater suspending power, environmentally acceptable and ease of handling. Oil-based fracturing fluids, more hazardous because of their flammability, are primarily used for water sensitive formation. Foam fracturing fluids are used in formations of low pressure and low per-*

*meability. They can reduce formation damage that could lead to blocked frac-*
*tures. But they cannot be loaded with high proppant concentration, and the*
*equipment cost is very high, uneconomical as compared to* ***aqueous*** *and oil-*
*based fracturing fluids.*

## Text

Until recently, the chemical composition of fracturing fluids was considered a trade secret and was not made public. This position has **fallen** increasingly **out of step with** public insistence that the community has the right to know what is being injected into the ground. Since 2010, voluntary disclosure has become the norm in most of the United States. The industry is also looking at ways to achieve the desired results without using potentially harmful chemicals. "Slick-water", made up of water, proppant, simple dragreducing polymers and biocide, has become increasingly popular as a fracturing fluid in the United States, though it needs to be pumped at high rates and can carry only very fine proppant. Attention is also being focused on reducing accidental surface spills, which most experts regard as a more significant risk of contamination to groundwater.

Chemical additives including biocides, surfactants, and scale **inhibitors** may also be incorporated into hydraulic fracturing fluids. Biocides such as **bromine**, methanol, or **naphthalene** eliminate bacterial organisms from **propagating** which can **clog** fractures and cause sliming in the wellbore. Surfactants such as **isopropanol** are used to manage the surface tension of the fracture fluid while **ethylene glycol** may be used to prevent **scale deposits** within the pipe. Other additives such as gelling agents, crosslinkers, and breakers may be used to control fluid viscosity and **impart** characteristics that control the fluid's capacity to carry or release proppant as needed.

Examples of the types of additives that may be present in fracturing fluids and their purpose are shown in Table1. The number, type, and concentration of chemical additives used in fracture treatments vary depending on specific well conditions and **formulations** are generally **proprietary** to the fracturing fluid producer. However, overall the concentration of additives in most slickwater fracturing fluids is relatively consistent 0. 5 percent to 2 percent with water making up 98 percent to 99. 5 percent of the solution.

**Table 5 – 1    Types and Purposes of Hydraulic Fracturing Fluid Additives**

| Additive Type | Purpose | Chemical Examples |
| --- | --- | --- |
| Acid | Cleans cement and drilling mud from perforation intervals prior to injection | Hydrochloric acid ( HCl 3% to 28% ) |
| Biocide | Inhibits growth of organisms that can contaminate methane gas and reduce fluid flow | Gluteraldehyde,2 – Bromo – 2nitro – ,2 – propanediol |
| Breaker | Reduces fluid viscosity to release proppant into fracture and aid fluid recovery | Peroxydisulfates |
| Clay Stabilizer | Prevents swelling and migration of formation clays which can reduce permeability | Salts( e. g, tertramethyl ammonium chloride, Potassium chloride( KCl) ) |
| Corrosion Inhibitor | Reduces rust formation on steel tubing, casing, tools, and tanks in acid-containing fracturing fluids | Methanol |
| Crosslinker | Crosslinking agents to enable increased fluid viscosity | Potassium hydroxide |
| Friction Reducer | Minimize friction to optimize fluid injection rate and pressure | Sodium acrylate-acrylamide copolymer |
| Gelling Agent | Increases fluid viscosity to deliver more proppant into fractures | Guar gum |
| Iron Control | Prevents precipitation of metal oxides which could plug the formation | Citric acid, thioglycolic acid |
| Proppant | "Props" fractures open to allow gas to flow more freely to the wellbore | Sand, sintered bauxite, ceramic beads |
| Scale Inhibitor | Prevents precipitation of carbonates and sulfates which could plug the formation | Ammonium chloride, ethylene glycol, polyacrylate |
| Surfactant | Reduces surface tension of fracturing fluids to aid fluid recovery | Methanol, isopropanol |

### *Foam Fracturing*

Foam fluids can replace slickwater formulations for hydraulic fracturing in some shales. The technology uses **inert gases**, typically nitrogen or carbon dioxide combined with foaming agents and water to create foam bubbles which carry and deposit proppant into shale fractures. Two key advantages of foam fracturing over high-pressure slickwater fracturing are reduction of formation damage that could lead to blocked fractures and substantial reduction of the volume of water that is needed for the fracturing.

Reducing the amount of water introduced into the formation decreases the potential for clays in the formation to swell, migrate, and plug fractures or the wellbore and reduces the potential volume of water that may remain trapped in the formation after flowback; improving production. Further, the amount of **process water** that must be transported to and from the well site for fracturing operations is decreased considerably. This translates to savings as the equipment and associated costs for water transport, storage equipment, treatment and disposal are reduced.

Foam fracturing technology finds use in lower permeability and lower pressure shales where particular consideration must be made to prevent induced formation damage and blockage from clay migration. The technology was initially developed decades ago and has been successfully utilized in the Appalachian Basin. However, testing in the early years of the Barnett Shale found the technology to **be incompatible with** portions of the play's **lithology** where brittle **quartz** is most successfully and economically fractured using high-pressure slickwater formulations. The technology continues to evolve with producers developing approaches to improve the stability and viscosity of the fracturing fluid, enhance transport characteristics for proppant, and reduce fluid loss and is gaining attention **as an alternative to** slickwater fracturing in potentially suitable plays, including the Fayetteville Shale in Arkansas and the Western Canadian Sedimentary Basin.

Another fracturing technique suitable for underpressurized, water-sensitive shales is nitrogen or carbon dioxide fracturing. This involves the use of a **non-aqueous fracturing fluid**, such as 100 percent nitrogen (or $CO_2$) as the carrier for specially engineered light-weight proppants. Eliminating the use of aqueous suspensions reduces capillary effects and formation damage associated with high-pressure hydrofracturing fluids. Formation **spalling** and generation of fine particles is also reduced due to the use of light-weight proppants, further enabling increased **conductivity** of the gas to the wellbore.

The technical breakthroughs discussed here have made finding, accessing, and producing gas from shale more efficient and effective than ever. Nowhere have these technologies been adopted more prolifically than in the major shale plays of the United States. This has resulted in a dramatic increase in the nation's production of shale gas over the past decade. In fact, shale gas enabled the United States, the world's biggest consumer of natural gas, to become the world's leading gas producer in 2009.

This remarkable success and the opinion of experts that shale gas exists in vast quantities across the continents have initiated a wave of international activity. Once largely ignored in favor of better understood and less challenging conventional gas sources, the prospect of energy security and independence has **piqued** international interest. Shale gas technology developed in the United States is rapidly becoming a global commodity through the efforts of major oilfield service companies. However, these technologies are not as mature as those used in conventional gas recovery and will continue to evolve to more efficiently address specific challenges presented by unique shale formations.

## Words and Expressions

| | |
|---|---|
| aqueous | 水的;水成的 |
| inhibitor | 抗化剂;禁制因素;约束者 |
| bromine | 溴(35号元素) |
| naphthalene | 萘球;卫生球;臭樟脑 |
| propagate | 扩散 |
| clog | 阻塞;妨碍 |
| isopropanol | 异丙醇 |
| impart | 授予 |
| formulation | 配方 |
| proprietary | 专利的;专有的 |
| consistent | 一致的 |
| lithology | 岩性学;结石学 |
| quartz | 石英 |
| spalling | 散裂 |
| conductivity | 传导性;传导率 |
| pique | 伤害……自尊心;激怒;产生;激起,引起……好奇心(兴趣) |

## Phrases and Expressions

| | |
|---|---|
| foam fluid | 泡沫压裂液 |
| fall out of step with | 与……不同步 |
| ethylene glycol | 乙二醇 |

| scale deposit | 结垢 |
| inert gas | 惰性气体 |
| process water | 工业用水 |
| be incompatible with | 与……不协调(不一致) |
| as an alternative to | 做为……替代品 |
| non-aqueous fracturing fluid | 无水压裂液 |

## Language Focus

1. Reducing the amount of water introduced into the formation decreases the potential for clays in the formation to swell, migrate, and plug fractures or the wellbore and reduces the potential volume of water that may remain trapped in the formation after flowback; improving production.

(参考译文:减少地层注水量可以降低黏土膨胀、运移及裂缝和井眼堵塞的可能性,减少返排后残留在地层中的水量,同时提高产量。)

本句中,动名词短语 Reducing the amount of water 作主语,decrease 和 reduce 为并列谓语动词,过去分词短语 introduced into the formation 作主语的后置定语;动词不定式符号 to 后接三个并列动词,分别为 swell,migrate 和 plug,并列连词 or 连接 fractures 和 wellbore;定语从句 that may remain trapped in the formation after flowback 修饰其前面的 the potential volume of water。

2. The technology continues to evolve with producers developing approaches to improve the stability and viscosity of the fracturing fluid, enhance transport characteristics for proppant, and reduce fluid loss and is gaining attention as an alternative to slickwater fracturing in potentially suitable plays, including the Fayetteville Shale in Arkansas and the Western Canadian Sedimentary Basin.

(参考译文:随着生产人员不断研究提高压裂液稳定性和粘度的方法,泡沫压裂技术也在不断进步,携带支撑剂的性能进一步提高,压裂液损失逐步降低。在条件适宜的气藏,包括阿肯色州的费耶特维尔页岩气藏和加拿大西部的沉积盆地气藏,用泡沫压裂液技术取代滑溜水压裂液技术受到越来越多的关注。)

本句是简单句。主谓结构是 The technology continues to evolve... and is gaining,主语带两个并列谓语;在第一个谓语中,不定式 to evolve 与后面的 enhance 和 reduce 并列,做目的状语;with producers developing 短语为独立主格结构,producers 和 developing 是逻辑上的主谓关系。Including 引导的现在

分词短语在句中作后置定语,修饰 potentially suitable plays。

3. Nowhere have these technologies been adopted more prolifically than in the major shale plays of the United States.

(参考译文:美国页岩气主产区使用这些技术生产效率最高。)

本句开头用了否定词 Nowhere,因此句子主谓半倒装,助动词 have 提到主语前。实际句子意思为 The major shale plays of United States has adopeted these technologies in the most prolifically way。

## 🔲 Reinforced Learning

### Ⅰ. Answer the following questions for a comprehension of the text.

1. What is regarded by most experts as a more significant risk of contamination to groundwater?

2. What is the concentration of additives in most slickwater fracturing fluids?

3. Describe the two key advantages of foam fracturing over high-pressure slickwater fracturing.

4. Which fracturing technique is suitable for underpressurized, water-sensitive shales?

5. What are the benefits of the use of light-weight proppants?

### Ⅱ. Multiple choice: choose the correct one from the alternative answers to give the exact meaning of the words.

1. Until recently, the chemical composition of fracturing fluids was considered a trade secret and was not made public.

A. compound　　　B. essay　　　C. community　　　D. ingredient

2. This position has fallen increasingly out of step with public insistence that the community has the right to know what is being injected into the ground.

A. constraint　　　B. claim　　　C. accessibility　　　D. pressure

3. The industry is also looking at ways to achieve the desired results without using potentially harmful chemicals.

A. firmly　　　B. latently　　　C. inevitably　　　D. commonly

4. Chemical additives including biocides, surfactants, and scale inhibitors may also be incorporated into hydraulic fracturing fluids.

A. reduced　　　B. combined　　C. progressed　　　D. defused

5. Biocides such as bromine, methanol, or naphthalene eliminate bacterial

organisms from <u>propagating</u> which can clog fractures and cause sliming in the wellbore.

A. circulating     B. broadcasting

C. transmitting     D. breeding

6. The number, type, and concentration of chemical additives used in fracture treatments vary depending on specific well conditions and <u>formulations</u> are generally proprietary to the fracturing fluid producer.

A. recipes   B. instructions C. explanations  D. agenda

7. The technology uses inert gases, typically nitrogen or carbon dioxide <u>combined</u> with foaming agents and water to create foam bubbles which carry and deposit proppant into shale fractures.

A. divided   B. mixed   C. connected   D. separated

8. However, testing in the early years of the Barnett Shale found the technology to be incompatible with portions of the play's lithology where <u>brittle</u> quartz is most successfully and economically fractured using high-pressure slick-water formulations.

A. complex   B. strong   C. durable   D. fragile

9. The technical <u>breakthroughs</u> discussed in this chapter have made finding, accessing, and producing gas from shale more efficient and effective than ever.

A. progress     B. important discoveries

C. transportation    D. reference

10. This <u>remarkable</u> success and the studied opinion of experts that shale gas exists in vast quantities across the continents have initiated a wave of international activity.

A. optional   B. removable C. notable   D. selectable

**III. Multiple choice: read the four suggested translations and choose the best answer.**

1. Biocides such as bromine, methanol, or naphthalene eliminate bacterial <u>organisms</u> from propagating which can clog fractures and cause sliming in the wellbore.

A. 生物   B. 物质   C. 营养    D. 群体

2. The number, type, and concentration of chemical additives used in fracture treatments vary depending on specific well conditions and formulations are generally <u>proprietary</u> to the fracturing fluid producer.

A. 未来的                          B. 光明的

C. 有希望的                        D. 业主所特有的

3. Reducing the amount of water introduced into the formation decreases the potential for clays in the formation to <u>swell</u>, migrate, and plug fractures or the wellbore and reduces the potential volume of water that may remain trapped in the formation after flowback.

A. 生长          B. 膨胀          C. 繁殖          D. 运动

4. However, testing in the early years of the Barnett Shale found the technology to be <u>incompatible</u> with portions of the play's lithology where brittle quartz is most successfully and economically fractured using high-pressure slick-water formulations.

A. 互助          B. 不相容     C. 促进          D. 抵抗

5. However, these technologies are not as <u>mature</u> as those used in conventional gas recovery and will continue to evolve to more efficiently address specific challenges presented by unique shale formations.

A. 长寿          B. 成熟          C. 完美          D. 健康

**Ⅳ. Put the following sentences into Chinese.**

1. This translates to savings as the equipment and associated costs for water transport, storage equipment, treatment and disposal are reduced.

2. Chemical additives including biocides, surfactants, and scale inhibitors may also be incorporated into hydraulic fracturing fluids.

3. The number, type, and concentration of chemical additives used in fracture treatments vary depending on specific well conditions and formulations are generally proprietary to the fracturing fluid producer.

4. Foam fracturing technology finds use in lower permeability; lower pressure shales where particular consideration must be made to prevent induced formation damage and blockage from clay migration.

5. Nowhere have these technologies been adopted more prolifically than in the major shale plays of theUnited States.

**Ⅴ. Put the following paragraphs into Chinese.**

1. Until recently, the chemical composition of fracturing fluids was considered a trade secret and was not made public. This position has fallen increasingly out of step with public insistence that the community has the right to know what

is being injected into the ground. Since 2010, voluntary disclosure has become the norm in most of the United States. The industry is also looking at ways to achieve the desired results without using potentially harmful chemicals.

2. Foam fluids can replace slickwater formulations for hydraulic fracturing in some shales. The technology uses inert gases, typically nitrogen or carbon dioxide combined with foaming agents and water to create foam bubbles which carry and deposit proppant into shale fractures. Two key advantages of foam fracturing over high-pressure slickwater fracturing are reduction of formation damage that could lead to blocked fractures and substantial reduction of the volume of water that is needed for the fracturing.

# 5.3　Waste Water Treatment and Disposal

## 🔲 Guidance to Reading

*Both conventional and unconventional oil and gas development can cause the potential for contamination of water supplies. Water could be contaminated by transportation **spills**, well casing leaks, leaks through fractured rock, hydrocarbons or chemicals migration from the produced zone into aquifers through the **intervening** rock, drilling site surface discharge, fracturing fluid flowback, coalbed methane **dewatering** and wastewater disposal. The greatest contamination risks came from wastewater disposal. Wastewater is dealt with in one of several ways, including but not limited to: disposal by underground injection, treatment followed by disposal to surface water bodies, or recycling for use in future hydraulic fracturing operations.*

## 🔲 Text

### *Waste water from hydraulic fracturing*

The treatment and disposal of waste water are critical issues for unconventional gas production – especially in the case of the large amounts of water customarily used for hydraulic fracturing. After being injected into the well, part of the fracturing fluid(which is often almost entirely water) is returned as flow-back in the days and weeks that follow. The total amount of fluid returned depends on the geology; for shale it can run from 20% to 50% of the input, the rest remaining bound to the clays in the shale rock. Flow-back water contains some of the chemicals used in the hydraulic fracturing process, together with metals, minerals and hydrocarbons **leached** from the reservoir rock. High levels of **salinity** are quite common and, in some reservoirs, the leached minerals can be weakly radioactive, requiring specific **precautions** at the surface. Flow-back returns (like waste water from drilling) requires secure storage on site, preferably fully contained in stable, **weather-proof** storage facilities as they do **pose a** potential **threat to** the local environment unless handled properly.

Once separated out, there are different options available for **dealing with** waste water from hydraulic fracturing. The optimal solution is to recycle it for future use and technologies are available to do this, although they do not always provide water **ready for** re-use for hydraulic fracturing on a **cost-effective** basis. A second option is to treat waste water at local industrial waste facilities capable of extracting the water and bringing it to a sufficient standard to enable it to be either discharged into local rivers or used in agriculture. Alternatively, where suitable geology exists, waste water can be injected into deep rock layers.

### Produced water from coalbed methane production

In the case of coalbed methane, additional water supplies are rarely required for the production process, but the satisfactory disposal of water that has been extracted from the well during the dewatering process is of critical importance. The produced water is usually either re-injected into isolated underground formations, discharged into existing **drainage** systems, sent to shallow ponds for evaporation or, once properly treated, used for irrigation or other productive uses. The appropriate disposal option depends on several factors, notably the quality of the water. Depending on the geology of the coal deposit and **hydrological** conditions, produced water can be very salty and **sodic** (containing high concentrations of sodium, calcium and magnesium) and can contain trace amounts of organic compounds, so it often requires treatment before it can be used for irrigation or other uses. Using **saline** water for irrigation can inhibit **germination** and plant growth, while excessively sodic water can change the physical properties of the soil, leading to poor drainage and crusting and **adversely** affecting crop yields.

The potential cost of water disposal depends on both the extent to which treatment is required and the volume of water produced. In practice, the total amount of water that must be removed from each well to allow gas to be produced varies considerably. It can be very large; for example, an estimated 65 cubic metres of water(17 000 gallons) are pumped from each coalbed methane well every day **on average** in the Powder River Basin in Montana and Wyoming. For the United States as a whole, it is estimated that, in 2008, more than 180 million cubic metres(47 billion gallons) of produced water were pumped out of coal seams, **equivalent to** the annual direct water consumption of the city of San Francisco.

**In principle**, produced water can be treated to any desired quality. This may be costly, but the treated water may have economic value for productive uses – as long as the cost of transporting the water is not excessive. The options for treatment and disposal of produced water and the market value of water **in the near vicinity** are often key factors in the economics of coalbed methane developments.

Many of the areas where coalbed methane is produced today, or where prospects for production are good, are **arid** or semi-arid and could benefit from additional freshwater supplies. For now, evaporation or discharge into drainage systems (in some cases, after treatment) are still the most common methods in North America(reuse of treated water is growing in Australia) because of the high cost of purifying the water for irrigation or reinjection into a deeper layer. In the United States, approximately 85 million cubic metres(22 billion gallons) of produced water, or about 45% of the total, were discharged to surface waters in 2008 with little or no treatment.

There is limited experience of assessing the actual environmental impacts of produced water from coalbed methane production. A recent study by the US National Research Council found that the eventual disposal or use of produced water can have both positive and negative impacts on soil, ecosystems, and the quality and quantity of surface water and groundwater. Although the study found no evidence of widespread negative effects, allowance must be made for the fact that the industry is relatively young and that few detailed investigations into local impacts have been carried out yet.

### *The risk of water contamination*

Significant concern has been expressed about the potential for contamination of water supplies, whether surface supplies, such as rivers or shallow freshwater aquifers, or deeper waters, as a result of all types of unconventional gas production. Water supplies can be contaminated from four main sources:

① Accidental spills of fluids or solids(drilling fluids, fracturing fluids, water and produced water, hydrocarbons and solid waste) at the surface.

② **Leakage** of fracturing fluids, saline water from deeper zones or hydrocarbons into a shallow aquifer through imperfect sealing of the cement column around the casing.

③ Leakage of hydrocarbons or chemicals from the producing zone to shallow aquifers through the rock between the two.

④ Discharge of insufficiently treated waste water into underground.

None of these hazards is specific to unconventional resources; they also exist in conventional developments, with or without hydraulic fracturing. However, unconventional developments occur at a scale that inevitably increases the risk of incidents occurring.

Public concern has focused on the third source of potential contamination, i. e. the possibility that hydrocarbons or chemicals might migrate from the produced zone into aquifers through the intervening rock. However, this may actually be the least significant of the hazards, at least in the case of shale gas and tight gas production; in some cases a focus on this risk may have diverted attention, including the time of regulators, away from other more pressing issues.

## 🔲 Words and Expressions

| | |
|---|---|
| spill | 溢出,流 |
| intervening | 插入的,中间的 |
| dewatering | 去水(作用),脱水(作用) |
| leach | 过滤,萃取,水浸 |
| salinity | 盐分,盐度;盐浓度;盐分 |
| precaution | 预防措施 |
| drainage | 排水;排水系统;污水 |
| hydrological | 水文学的 |
| sodic | 钠的,含钠的 |
| saline | 含盐的,咸的 |
| germination | 萌芽,发生;萌发;生芽;催芽 |
| adversely | 不利地 |
| arid | 干旱的,干燥的;贫瘠的,荒芜的, |
| leakage | 泄漏 |

## Phrases and Expressions

| | |
|---|---|
| weather-proof | 防风雨的 |
| pose a threat to | 对……构成威胁 |
| deal with | 对待(对付)某人,应付(处理)问题、事务等 |
| be ready for | 为……做好准备 |
| cost-effective | 有成本效益的,划算的 |
| on average | 平均 |
| equivalent to | 相当于 |
| in principle | 原则上;基本上 |
| in the near vicinity | 在附近 |

## Language Focus

1. The optimal solution is to recycle it for future use and technologies are available to do this, although they do not always provide water ready for re-use for hydraulic fracturing on a cost-effective basis.

(参考译文:最佳方法是将其回收留待日后利用,这在技术上完全可行,不过出于经济方面的考虑,还不能随时都有水循环用于水力压裂。)

本句是并列复合句,主体部分是 and 连接的并列句,这两个句子都是主、系、表结构。although 引导让步状语从句,ready for 是形容词短语做 water 的后置定语。

2. Depending on the geology of the coal deposit and hydrological conditions, produced water can be very salty and sodic (containing high concentrations of sodium, calcium and magnesium) and can contain trace amounts of organic compounds, so it often requires treatment before it can be used for irrigation or other uses.

[参考译文:根据不同的煤层沉积环境和水文条件,采出水可能盐、钠含量很高(含高浓度的钠、钙、镁),另外还可能含有微量有机化合物,因此通常需要处理后才能用于农业灌溉等。]

本句为复合句,开头的现在分词短语 Depending on 做状语,主句有两个并列的谓语动词:can be 和 can contain;从句是 so 引导的结果状语,而这个从句又含有一个 before 引导的时间状语从句。

3. Using saline water for irrigation can inhibit germination and plant growth, while excessively sodic water can change the physical properties of the soil, lead-

ing to poor drainage and crusting and adversely affecting crop yields.

（参考译文：使用盐水灌溉会抑制植物发芽与生长，而水中钠含量过高会改变土壤的物理性质，导致排水不良及地表结壳，最终影响农作物的产量。）

本句中，动名词短语 Using saline water for irrigation 作主句主语，while 引导从句，和前面的句子进行对比；现在分词短语 leading to... 作结果状语。

## 凸 Reinforced Learning

### I. Answer the following questions for a comprehension of the text.

1. What doesthe flow-back water consist of ?

2. Describe the optimal solution of dealing with waste water from hydraulic fracturing.

3. What influence would excessively sodic water bring?

4. Why isevaporation or discharge into drainage systems still the most common methods in North America, when dealing with the produced water?

5. Whatdoes the potential cost of water disposal depend on?

### II. Multiple choice: choose the correct one from the alternative answers to give the exact meaning of the words.

1. The treatment and disposal of waste water are critical issues for unconventional gas production – especially in the case of the large amounts of water customarily used for hydraulic fracturing.

A. frequently     B. usually     C. continuously     D. soon

2. Flow-back water contains some of the chemicals used in the hydraulic fracturing process, together with metals, minerals and hydrocarbons leached from the reservoir rock.

A. contained     B. reached     C. filtered     D. absorbed

3. Flow-back returns (like waste water from drilling) requires secure storage on site, preferably fully contained in stable, weather-proof storage facilities as they do pose a potential threat to the local environment unless handled properly.

A. begin     B. achieve     C. carry     D. present

4. A second option is to treat waste water at local industrial waste facilities capable of extracting the water and bringing it to a sufficient standard to enable it to be either discharged into local rivers or used in agriculture.

A. opinion　　　　B. alternative　　C. action　　　　D. decision

5. The produced water is usually either re-injected into <u>isolated</u> underground formations, discharged into existing drainage systems, sent to shallow ponds for evaporation or, once properly treated, used for irrigation or other productive uses.

A. soft　　　　　B. smooth　　　　C. amused　　　　D. separated

6. Using saline water for irrigation can inhibit germination and plant growth, while <u>excessively</u> sodic water can change the physical properties of the soil, leading to poor drainage and crusting and adversely affecting crop yields.

A. fully　　　　　B. enough　　　　C. overly　　　　　D. hurriedly

7. For the United States as a whole, it is estimated that, in 2008, more than 180 million cubic metres (47 billion gallons) of produced water were pumped out of coal seams, <u>equivalent</u> to the annual direct water consumption of the city of San Francisco..

A. more　　　　　B. same　　　　　C. higher　　　　　D. lower

8. A recent study by the US National Research Council found that the eventual disposal or use of produced water can have both positive and <u>negative</u> impacts on soil, ecosystems, and the quality and quantity of surface water and groundwater.

A. complex　　　　B. bad　　　　　C. durable　　　　D. fragile

9. However, as noted, unconventional developments occur at a scale that <u>inevitably</u> increases the risk of incidents occurring.

A. normally　　　　B. necessarily　　C. frankly　　　　D. in time

10. However, this may actually be the least significant of the <u>hazards</u>, at least in the case of shale gas and tight gas production; in some cases a focus on this risk may have diverted attention, including the time of regulators, away from other more pressing issues.

A. harm　　　　　B. benefits　　　　C. interests　　　　D. selections

**III. Multiple choice: read the four suggested translations and choose the best answer.**

1. High levels of salinity are quite common and, in some reservoirs, the leached minerals can be weakly <u>radioactive,</u> requiring specific precautions at the surface.

A. 可告知的　　　B. 传播性的　　　C. 放射性的　　　　D. 易感染的

2. The optimal solution is to recycle it for future use and technologies are available to do this, although they do not always provide water ready for re-use for hydraulic fracturing on a cost-effective basis.

    A. 方案            B. 解决办法    C. 进程         D. 处理方式

3. For now, evaporation or discharge into drainage systems( in some cases, after treatment) are still the most common methods in North America( re-use of treated water is growing in Australia) because of the high cost of purifying the water for irrigation or reinjection into a deeper layer.

    A. 生长            B. 膨胀         C. 蒸发         D. 运动

4. There is limited experience of assessing the actual environmental impacts of produced water from coalbed methane production.

    A. 攻击            B. 发现         C. 评估          D. 探索

5. Although the study found no evidence of widespread negative effects, allowance must be made for the fact that the industry is relatively young and that few detailed investigations into local impacts have been carried out yet.

    A. 必须赞美       B. 必须肯定    C. 必须考虑到    D. 必须允许

**Ⅳ. Put the following sentences into Chinese.**

1. The treatment and disposal of waste water are critical issues for unconventional gas production – especially in the case of the large amounts of water customarily used for hydraulic fracturing.

2. In the case of coalbed methane, additional water supplies are rarely required for the production process, but the satisfactory disposal of water that has been extracted from the well during the dewatering process is of critical importance.

3. The produced water is usually either re-injected into isolated underground formations, discharged into existing drainage systems, sent to shallow ponds for evaporation or, once properly treated, used for irrigation or other productive uses.

4. This may be costly, but the treated water may have economic value for productive uses – as long as the cost of transporting the water is not excessive.

5. A recent study by the US National Research Council found that the eventual disposal or use of produced water can have both positive and negative impacts on soil, ecosystems, and the quality and quantity of surface water and groundwater.

**V. Put the following paragraphs into Chinese**

1. The appropriate disposal option depends on several factors, notably the quality of the water. Depending on the geology of the coal deposit and hydrological conditions, produced water can be very salty and sodic and can contain trace amounts of organic compounds, so it often requires treatment before it can be used for irrigation or other uses. Using saline water for irrigation can inhibit germination and plant growth, while excessively sodic water can change the physical properties of the soil, leading to poor drainage and crusting and adversely affecting crop yields.

2. Public concern has focused on the third source of potential contamination, i. e. the possibility that hydrocarbons or chemicals might migrate from the produced zone into aquifers through the intervening rock. However, this may actually be the least significant of the hazards, at least in the case of shale gas and tight gas production; in some cases a focus on this risk may have diverted attention, including the time of regulators, away from other more pressing issues.

# 5.4 Greenhouse-gas Emissions

## Guidance to Reading

*The primary greenhouse gases in the Earth's atmosphere are water vapor, carbon dioxide, methane, nitrous oxide and ozone. Since the beginning of the Industrial Revolution, the burning of fossil fuels and extensive clearing of native forests has contributed to a 40% increase in the atmospheric concentration of carbon dioxide. Methane is the second most prevalent greenhouse gas emitted from human activities. Methane is emitted by natural sources such as wetlands, as well as human activities such as leakage from natural gas systems and the raising of livestock. Methane's lifetime in the atmosphere is much shorter than carbon dioxide, but the comparative impact of methane on climate change is over 20 times greater than carbon dioxide over a 100-year period.*

## Text

Shale gas and tight gas have higher production-related greenhouse-gas emissions than conventional gas. This **stems from** two effects:

① *More wells and more hydraulic fracturing are needed per cubic metre of gas produced*. These operations use energy, typically coming from **diesel motors**, leading to higher $CO_2$ emissions per unit of useful energy produced.

② *More venting or flaring during well completion*. The flow-back phase after hydraulic fracturing represents a larger percentage of the total recovery per well (because of more hydraulic fracturing, the flow-back takes longer and the total recovery per well is typically smaller due to the low permeability of the rock).

We have previously released estimates of these effects both in the case of flaring and for venting during flow-back, based on EPA data, in order to see what difference these practices make. In the case of flaring, total well-to-burner emissions are estimated to be 3.5% higher than for conventional gas, but this figure rises to 12% if the gas is vented. Eliminating venting, minimising flaring and recovering and selling the gas produced during flow-back, **in line with** the Golden Rules, would reduce emissions below the lower figure given here.

Similar concerns about emissions attach to coalbed methane production, where significant volumes of methane can be vented into the atmosphere during the transition phase from dewatering to gas production and, where hydraulic fracturing is applied, during the well completion phase. Careful management of drilling, fracturing and production operations is essential to keep such emissions to a minimum. This requires specialised equipment to separate gas from the produced water or fracturing fluids before injecting it into a gasgathering system or into temporary storage. If this is not possible for technical, logistical or economic reasons, it is preferable that the gas should be flared rather than vented for safety reasons and because the global-warming effect is considerably less.

The general issue of greenhouse-gas emissions from the production, transportation and use of natural gas, as well as the additional emissions from unconventional gas compared with conventional gas, has been the subject of some **controversy**. It is argued that emissions from using natural gas as a source of primary energy have been significantly underestimated, particularly for unconventional gas. It has even been argued that full life-cycle emissions from unconventional gas can be higher than from coal. The main issue revolves around methane emissions not only during production, but also during transportation and use of natural gas.

Methane is a more **potent** greenhouse gas than $CO_2$ but has a shorter lifetime in the atmosphere — a half-life of about fifteen years, **versus** more than 150 years for $CO_2$. As a result, there are different possible ways to compare the effect of methane and $CO_2$ on global warming. One way is to evaluate the Global Warming Potential(GWP) of methane, compared to $CO_2$, averaged over 100 years. The 4th assessment report of the **IPCC** gives a value of 25 for this 100-years GWP, revised up from their previous estimate of 21. This value is relevant when looking at the long-term relative benefits of eliminating a temporary source of methane emissions versus a $CO_2$ source.

Averaged over 20 years, the GWP, estimated by the IPCC, is 72. This figure can be argued to be more relevant to the evaluation of the significance of methane emissions in the next two or three decades, which will be the most critical to determine whether the world can still reach the objective of limiting the long-term increase in average surface temperatures to 2 degrees Celsius. Moreover, some scientists have argued that interactions of methane with **aerosols** reinforce the GWP of methane, possibly bringing it to a higher level.

Such higher values would, of course, have **implications** not only for methane emissions from the gas chain but also for all other methane emissions, from **livestock**, landfills, **rice paddies** and other agricultural sources, as well as from natural sources.

Methane emissions along the gas value chain(whether conventional or unconventional) come from four main sources:

① *Intentional venting of gas for safety or economic reasons.* Venting during well completions falls into this **category**, but venting can also take place as part of equipment maintenance operations.

② *Fugitive emissions.* These might be leaks in pipelines, valves or seals, whether accidental(e. g. corrosion in pipelines) or built into the equipment design (e. g. rotating seals, open tanks).

③ *Incidents involving rupture of confining equipment*(pipelines, pressurised tanks, well isolation).

④ *Incomplete burning.* The effectiveness of gas burning in gas flares varies according to wind and other conditions and is typically no better than 98%. (A similar effect can be seen when starting a gas stove:it can take a few seconds before a steady flame is established).

By their very nature, these emissions are difficult to quantify. Most estimates are based on **emission factors** for various parts of the chain(wells, various equipment, pipelines and so on), derived from studies conducted in the United States by the EPA and the Gas Research Institute in the 1990s. It is **by no means** clear that these studies give a good indication for emissions in other parts of the world, or for the possible evolution of methane emissions in the future. Estimates of methane emissions from the gas chain at the global level vary between 1% and 8% of produced natural gas volumes. The most comprehensive **projections** of future emissions, from the EPA( US EPA, 2011), assume no change in emission factors, for want of a better approach, and project a 26% increase in methane emissions from the oil and gas industry between 2010 and 2030.

Different **assumptions** about the level and impact of methane emissions can have a profound effect on the **perception** of gas as a "cleaner" fossil fuel. The advantage of gas over coal is widely accepted due to its lower $CO_2$ emissions per unit of energy from combustion; but it is clear that more pessimistic assumptions can make gas a worse greenhouse-gas emitter than coal. It is very important that additional scientific work should **pinpoint** the most relevant GWP value and that efforts are redoubled to measure methane emissions more systematically.

## Words and Expressions

| | |
|---|---|
| potent | 有效的,强有力的 |
| versus | (表示两队或双方对阵)对;(比较两种不同想法等)与……相对; |
| aerosols | 气溶胶(aerosol 的复数);喷雾剂;(气体中的)浮粒 |
| implication | 隐含意义;寓意 |
| livestock | 家畜,牲畜 |
| category | 类型,部门,种类 |
| fugitive | 逃亡的;难以捉摸的 |
| rupture | 破裂 |
| projection | 预测 |
| assumption | 假设 |
| perception | 知觉;觉察(力);观念 |
| pinpoint | 确定,准确地指出;精准定位 |
| controversy | 公开辩论;论战;争议 |

## Phrases and Expressions

| | |
|---|---|
| stem from | 起源于,由……造成 |
| diesel motors | 柴油发动机 |
| in line with | 跟……一致,符合;本着 |
| IPCC | 政府间气候变化专门委员会(Intergovernmental Panel on Climate Change) |
| rice paddy | 稻田;水田 |
| emission factor | 排放系数 |
| by no means | 决不 |

## Language Focus

1. Similar concerns about emissions attach to coalbed methane production, where significant volumes of methane can be vented into the atmosphere during the transition phase from dewatering to gas production and, where hydraulic fracturing is applied, during the well completion phase.

(参考译文:煤层气开发中温室气体排放问题同样受到关注,完井时的水力压裂以及排水到煤层气生产过程中,都会有大量甲烷排入大气中。)

本句为复合句,主句是 simialar concerns attach to coalbed methane production,后跟两个 where 引导的非限制性定语从句,均修饰先行词 coalbed methane production。

2. If this is not possible for technical, logistical or economic reasons, it is preferable that the gas should be flared rather than vented for safety reasons and because the global-warming effect is considerably less.

(参考译文:如果出于技术、后勤或者经济上的原因难以做到的话,最好将气体烧掉而不是将其排出,因为这样做安全且产生的温室效应要小得多。)

本句主句是 it is preferable that...,it 是形式主语,that 从句是真正的主语,rather than 表示比较,and 连接了两个原因状语,一个是介词短语 for safety reasons,另一个是 because 从句。if 引导条件状语从句,for...reasons 表示原因。

3. This figure can be argued to be more relevant to the evaluation of the significance of methane emissions in the next two or three decades, which will be the most critical to determine whether the world can still reach the objective of limit-

ing the long-term increase in average surface temperatures to 2 degrees Celsius.

（参考译文：可以说这一数字更接近将来二三十年甲烷的大量排放估值，是决定世界能否达到控制地表平均温度增长不超过2摄氏度这一目标的关键。）

本句中，which 引导非限制性定语从句，修饰 the evaluation of the significance of methane emissions；而在目的状语 to determine... 中，whether 引导的从句做 determine 的宾语。

## 🔲 Reinforced Learning

### Ⅰ. Answer the following questions for a comprehension of the text.

1. Why does unconventional gas production lead to higher greenhouse-gas emissions than conventional gas?

2. What is required tobe done before the gas is injected into a gasgathering system in coalbed methane production?

3. Please give one of the ways to compare the effect of methane and carbon dioxide on global warming.

4. What else can lead to methane emissions besides gas production?

5. Why can different assumptions about the level and impact of methane emissions have a profound influence on the perception of gas as a "cleaner" fossil fuel?

### Ⅱ. Multiple choice: choose the correct one from the alternative answers to give the exact meaning of the words.

1. We have previously released estimates of these effects both in the case of flaring and for venting during flow-back, based on EPA data, in order to see what difference these practices make.

A. accidentally　　B. formerly　　C. antecedently　　D. truly

2. This requires specialised equipment to separate gas from the produced water( and fracturing fluids)before injecting it into a gasgathering system( or into temporary storage).

A. contained　　B. dedicated　　C. filtered　　D. absorbed

3. This value is relevant when looking at the long-term relative benefits of eliminating a temporary source of methane emissions versus a $CO_2$ source.

A. completed　　B. firm　　C. eventual　　D. momentary

4. A second option is to treat waste water at local industrial waste facilities capable of extracting the water and bringing it to a sufficient standard to enable it to be either <u>discharged</u> into local rivers or used in agriculture.

A. outpoured　　　B. damaged　　　C. disclosed　　　D. defused

5. Such higher values would, of course, have <u>implications</u> not only for methane emissions from the gas chain but also for all other methane emissions, from livestock, landfills, rice paddies and other agricultural sources, as well as from natural sources.

A. applications　　　　　　　　B. formations

C. associations　　　　　　　　D. formulations

6. Venting during well completions falls into this category, but venting can also take place as part of equipment <u>maintenance</u> operations.

A. construction　　B. exploration　　C. machine　　　D. upkeep

7. Most estimates are based on emission factors for various parts of the chain( wells, various equipment, pipelines and so on), <u>derived</u> from studies conducted in the United States by the EPA and the Gas Research Institute in the 1990s.

A. reached　　　B. discussed　　　C. talked　　　D. coming

8. It is by no means clear that these studies give a good <u>indication</u> for emissions in other parts of the world, or for the possible evolution of methane emissions in the future.

A. revolution　　　B. discussion　　C. hint　　　D. foundation

9. The most comprehensive projections of future emissions, from the EPA, <u>assume</u> no change in emission factors, for want of a better approach, and project a 26% increase in methane emissions from the oil and gas industry between 2010 and 2030.

A. believe　　　B. boast　　　C. presume　　　D. scare

10. Different assumptions about the level and impact of methane emissions can have a <u>profound</u> effect on the perception of gas as a"cleaner"fossil fuel.

A. timely　　　B. superficial　　C. deep　　　D. occasional

**Ⅲ. Multiple choice : read the four suggested translations and choose the best answer.**

1. Similar concerns about emissions attach to coalbed methane production, where significant volumes of methane can be vented into the atmosphere during

the <u>transition</u> phase from dewatering to gas production and, where hydraulic fracturing is applied, during the well completion phase.

  A. 计划    B. 决定    C. 过渡    D. 实践

  2. Incidents involving <u>rupture</u> of confining equipment( pipelines, pressurised tanks, well isolation).

  A. 推导    B. 改进    C. 修缮    D. 破损

  3. The effectiveness of gas burning in gas flares varies according to wind and other conditions and is typically no better than 98%. ( A similar effect can be seen when starting a gas <u>stove</u>: it can take a few seconds before a steady flame is established).

  A. 地点    B. 厨房    C. 火炉    D. 现场

  4. The advantage of gas is widely accepted due to its lower $CO_2$ emissions per unit of energy from <u>combustion</u>.

  A. 漂浮    B. 喷射    C. 泄流    D. 燃烧

  5. It is very important that additional scientific work should pinpoint the most relevant GWP value and that efforts are redoubled to measure methane emissions more <u>systematically</u>.

  A. 系统地    B. 机械地    C. 全部地    D. 个别地

**Ⅳ. Put the following sentences into Chinese.**

  1. We have previously released estimates of these effects both in the case of flaring and for venting during flow-back, based on EPA data, in order to see what difference these practices make.

  2. Eliminating venting, minimising flaring and recovering and selling the gas produced during flow-back, in line with the Golden Rules, would reduce emissions below the lower figure given here.

  3. It has even been argued that full life-cycle emissions from unconventional gas can be higher than from coal.

  4. This value is relevant when looking at the long-term relative benefits of eliminating a temporary source of methane emissions versus a $CO_2$ source.

  5. It is by no means clear that these studies give a good indication for emissions in other parts of the world, or for the possible evolution of methane emissions in the future.

## V. Put the following paragraphs into Chinese.

1. Careful management of drilling, fracturing and production operations is essential to keep such emissions to a minimum. This requires specialised equipment to separate gas from the produced water or fracturing fluids before injecting it into a gasgathering system or into temporary storage. If this is not possible for technical, logistical or economic reasons, it is preferable that the gas should be flared rather than vented for safety reasons and because the global-warming effect is considerably less.

2. The general issue of greenhouse-gas emissions from the production, transportation and use of natural gas, as well as the additional emissions from unconventional gas compared with conventional gas, has been the subject of some controversy. It is argued that emissions from using natural gas as a source of primary energy have been significantly underestimated, particularly for unconventional gas. It has even been argued that full life-cycle emissions from unconventional gas can be higher than from coal. The main issue revolves around methane emissions not only during production, but also during transportation and use of natural gas.

# Chapter 6　Unconventional Oil and Gas Market

## 6.1　Production Costs and Price of Natural Gas

### 📑 Guidance to Reading

*Historically, international trade in gas was quite limited, as gas was produced and consumed locally. Pricing mechanisms ranged from regulated prices set by governments, prices indexed to competing fuels, specifically oil products, or spot market pricing in competitive markets. Due to different geography or market conditions, regions in the world produce gases at different costs and hence set different prices. On the basis of various gas producing costs, one can estimate a "**break-even cost**", the cost of output at which the revenue received by the business is exactly equal to the cost of its production. A breakeven approach can benefit cost-effectiveness analysis as well as price setting.*

### 📑 Text

The costs of developing and producing unconventional gas are made up of several elements: capital costs, operational costs, transportation costs, and taxes and **royalties**.

Capital costs, often called finding and development costs, are usually dominated by the costs of constructing wells. Shale gas wells do cost more than conventional gas wells in the same conditions, because of the additional costs of multistage hydraulic fracturing; the same consideration applies to tight gas wells, for the same reason. Coalbed methane wells have so far been relatively cheap, compared with conventional gas wells, because production has been at shallow depths in regions with well-developed markets. Operational costs, also called **lifting costs**, are those variable costs that are directly linked to the production activity: they may differ according to local conditions. The cost of bringing gas to market is distance-dependent and is **identical** for conventional and unconventional gas.

The final element, taxes and royalties, varies greatly between **jurisdictions**;

in addition to a profit tax component, it very often includes fixed or production-related taxes (paid to governments) and/or royalties (paid to the resource owner, which may or may not be governments). Countries or regions that have higher capital and operating costs, due to their geography or market conditions, often create a more attractive **fiscal regime** in order to attract investment. This can go as far as offering subsidies: China provides **subsidies** for coalbed methane and shale gas production.

On the basis of these costs, one can estimate a "break-even cost", or "**supply cost**", the market value required to provide an adequate real return on capital for a new project. Lifting costs, transport costs, and taxes and royalties are usually directly expressed in US dollars per unit of gas produced. The significance of capital costs is very dependent on the amount of gas recovered per well. This also varies greatly: the best shale gas wells in the United States are reported to have **Estimated Ultimate Recovery** (EUR) of 150 to 300 million cubic metres (mcm) (5 to 10 billion cubic feet [bcf]); but many shale gas wells have EUR that is 10 or 100 times less. The average EUR varies from one shale to another, but also depends on the experience of the industry in a given shale: with time, the industry **optimises** the technologies used and extracts more gas from each well. Outside the United States, there is essentially no experience so far, but drilling longer horizontal wells should help improve EUR per well (in many jurisdictions in the United States, horizontal well length is limited by **acreage** unit size regulations).

It follows from the discussion of costs that the break-even costs for gas can vary greatly from one location to the next, or within a single country. For example in the United States, break-even costs for dry gas wells probably range from $ 5/MBtu to $ 7/MBtu; gas containing liquids has a lower (gas) break-even cost, which can be as low as $ 3/MBtu, as the liquids add considerable value for a small increase in costs (associated gas from wells producing predominantly oil can have an even lower break-even cost).

Since conventional gas resources are already fairly **depleted** onshore and most future conventional gas production will therefore come from more expensive offshore locations, the range of break-even costs for conventional and unconventional gas in the United States is fairly similar.

In Europe, the costs of production are expected to be about 50% higher, with a range of break-even costs between $ 5/MBtu and $ 10/MBtu. Conventional and unconventional gases are expected to be in the same range, as conventional resources are depleted and new projects are moving to the more expensive Norwegian Arctic. China has a cost structure similar to that of the United States, but shale reservoirs there tend to be deeper and more geologically complex; similarly, coalbed methane reservoirs in China tend to be in **remote** locations, so we estimate the break-even cost range to be **intermediate** between that of the United States and that of Europe – from $ 4/MBtu to $ 8/MBtu (although there are production subsidies in place that can bring this figure down). This estimate for China applies to both conventional and unconventional gas, as the easy conventional gas is depleting and production is moving to offshore or more remote regions. In countries that have large, relatively easy, remaining conventional gas, such as the Middle East, with break-even costs of less than $ 2/MBtu, the break-even cost range for unconventional gas is expected to be higher (similar to that for unconventional gas in the United States).

North America is the region where the unconventional gas industry has grown most rapidly and, unsurprisingly, is also the region where the impact on markets and prices has thus far been greatest. Historically low prices are being obtained for natural gas, relative to other energy forms such as oil. More surprisingly, given the relative isolation of North American markets from other major gas-using regions, this development has already had profound international impacts. These have arisen because North America has become almost selfsufficient in gas, whereas many LNG investments in the decade 2000 to 2010 were made in the expectation that the North American region would be a substantial net LNG importer.

**Import infrastructure** in excess of 100 bcm was built in the United States alone in this period, with matching LNG supply investments in major producers, such as Qatar. However, in 2011, net LNG imports to North America were less than 20 bcm, out of a total market exceeding 850 bcm: 8 bcm into the United States and 9 bcm into Mexico and Canada. Hence, major quantities of LNG supply became available for other global markets, including Asia and Europe.

At present, gas prices are set freely in several markets, including North America, the United Kingdom and, to a somewhat lesser extent, Australia, an approach known as **gas-to-gas competition**. However, much of the gas traded across borders in **the Asia-Pacific region** is sold under long-term contracts, with linkages to the price of oil or **refined** products. Prices in continental Europe are **pre-**

**dominantly** oil-linked, though in recent years a mixture of the two systems (and many variations in between) has emerged, with **oil-indexed prices** co-existing – often uneasily – with prices set by gas-to-gas competition. Energy security, **geopolitics**, and the shift to greener forms of fuel that will be critical for combating climate change will depend on how gas pricing evolves.

## Words and Expressions

| | |
|---|---|
| royalties | 矿区土地使用费(由采矿或石油公司等付给土地所有人) |
| identical | |
| jurisdiction | 裁判权;管辖区域;管辖范围 |
| optimise | 使优化;充分利用 |
| subsidies | 补助金 |
| acreage | 面积;英亩数 |
| deplete | 枯竭,耗尽 |
| remote | (时间上)遥远的;远离的;远程的;微小的 |
| intermediate | 中间的;中级的 |
| refined | 精炼的;精制的;经过改良的 |
| predominantly | 占主导地位地;显著地;占优势地 |
| geopolitics | 地缘政治 |

## Phrases and Expressions

| | |
|---|---|
| break-even cost | 无盈亏成本 |
| lifting cost | 采收成本 |
| fiscal regime | 财政制度 |
| supply cost | 供应成本 |
| Estimated Ultimate Recovery | 预计最终采收率 |
| import infrastructure | 进口基础设施 |
| gas-to-gas competition | 气之间的竞争 |
| the Asia-Pacific region | 亚太地区 |
| oil-indexed prices | 与油价挂钩的价格 |

## Language Focus

1. Countries or regions that have higher capital and operating costs, due to their geography or market conditions, often create a more attractive fiscal regime in order to attract investment.

（参考译文：由于地理或市场条件不同，资本和运营成本较高的国家或地区，通常会建立更具吸引力的财政制度吸引投资，有时甚至可能提供补贴：中国对煤层气和页岩气生产就采取了补贴政策。）

本句主句主谓结构为 Countries or regions... often create..., that 引导定语从句，修饰 countries or regions。due to 介词短语做状语，意思是"由于……"。

2. Since conventional gas resources are already fairly depleted onshore and most future conventional gas production will therefore come from more expensive offshore locations, the range of break-even costs for conventional and unconventional gas in the United States is fairly similar.

（参考译文：由于陆上常规天然气资源已接近枯竭，未来常规天然气产量将来自成本更高的海上，因此，在美国常规和非常规天然气盈亏平衡成本范围基本上是一致的。）

本句中，since 引导原因状语从句，由 and 连接两个并列的原因。

3. In countries that have large, relatively easy, remaining conventional gas, such as the Middle East, with break-even costs of less than ＄2/MBtu, the break-even cost range for unconventional gas is expected to be higher.

［参考译文：在那些仍然拥有大量相对容易开采的常规天然气国家，如中东地区国家，盈亏平衡成本还不到2美元/MBtu，但其非常规天然气盈亏平衡成本会相对高一些（类似于美国的非常规天然气）。］

本句开头 in countries 是地点状语，后面跟有两个定语修饰 countries，一个是 that 引导的定语从句，另一个是 with 介词短语；主句是 he break-even cost range for unconventional gas is expected to be higher。

## Reinforced Learning

### I. Answer the following questions for a comprehension of the text.

1. What are the costs of developing and producing unconventional gas made up of?

2. Why have coalbed methane wells so far been relatively cheap compared

with conventional gas wells?

3. Why does a more attractive fiscal regime need to be created in gas development?

4. Why were many LNG investments made in the decade 2000 to 2010 in North America?

5. Please describe the socalled "gas-to-gas competition".

**II. Multiple choice: choose the correct one from the alternative answers to give the exact meaning of the words.**

1. The final element, taxes and royalties, varies greatly between jurisdictions; in addition to a profit tax component, it very often includes fixed or production-related taxes (paid to governments) and/or royalties.

A. payment　　　B. salary　　　C. earnings　　　D. wage

2. This can go as far as offering subsidies: China provides subsidies for coalbed methane and shale gas production.

A. substitutes　　B. subsidiaries　　C. allowances　　D. substances

3. China has a cost structure similar to that of the United States, but shale reservoirs there tend to be deeper and more geologically complex; similarly, coalbed methane reservoirs in China tend to be in remote locations, so we estimate the break-even cost range to be intermediate between that of the United States and that of Europe-from $4/MBtu to $8/MBtu.

A. completed　　B. underground　　C. eventual　　　D. medium

4. However, much of the gas traded across borders in the Asia-Pacific region is sold under long-term contracts, with linkages to the price of oil or refined products.

A. connections　　B. packages　　　C. solutions　　　D. conditions

5. Historically low prices are being obtained for natural gas, relative to other energy forms such as oil. More surprisingly, given the relative isolation of North American markets from other major gas-using regions, this development has already had profound international impacts.

A. found　　　B. acquired　　　C. discussed　　　D. talked

6. These have arisen becauseNorth America has become almost selfsufficient in gas, whereas many LNG investments in the decade 2000 to 2010 were made in the expectation that the North American region would be a substantial net LNG importer.

A. sustaining        B. significant        C. sigmficant        D. tiny

7. Shale gas wells do cost more than conventional gas wells in the same conditions, because of the <u>additional</u> costs of multistage hydraulic fracturing; the same consideration applies to tight gas wells, for the same reason.

A. total        B. sufficient        C. extra        D. previous

8. On the basis of these costs, one can estimate a "break-even cost", or "supply cost", the market value required to provide an <u>adequate</u> real return on capital for a new project.

A. good        B. enough        C. perfect        D. loyal

9. It follows from the discussion of costs that the break-even costs for gas can vary greatly from one <u>location</u> to the next, or within a single country.

A. work        B. sound        C. place        D. country

10. Conventional and unconventional gases are expected to be in the same range, as conventional resources are depleted and new <u>projects</u> are moving to the more expensive Norwegian Arctic.

A. substances        B. funds        C. programs        D. constructions

**III. Multiple choice: read the four suggested translations and choose the best answer.**

1. The final element, <u>taxes</u> and royalties, varies greatly between jurisdictions; in addition to a profit tax component, it very often includes fixed or production-related taxes(paid to governments) and/or royalties.

A. 费用        B. 税额        C. 收入        D. 佣金

2. Outside the United States, there is essentially no experience so far, but drilling longer <u>horizontal</u> wells should help improve EUR per well.

A. 直线的        B. 垂直的        C. 水平的        D. 曲线的

3. Prices in <u>continental</u> Europe are predominantly oil-linked, though in recent years a mixture of the two systems(and many variations in between) has emerged, with oil-indexed prices co-existing – often uneasily – with prices set by gas-to-gas competition.

A. 地球的        B. 洲的        C. 大陆的        D. 地点的

4. These have arisen because North America has become almost selfsufficient in gas, whereas many LNG investments in the decade 2000 to 2010 were made in the expectation that the North American region would be a substantial net LNG <u>importer</u>.

A. 制造商　　B. 出口商　　C. 进口商　　D. 销售者

5. Countries or regions that have higher capital and operating costs, due to their geography or market conditions, often create a more attractive <u>fiscal regime</u> in order to attract investment.

A. 财政制度　　B. 表面政权　　C. 社会制度　　D. 管理体制

**Ⅳ. Put the following sentences into Chinese.**

1. Operational costs, also called lifting costs, are those variable costs that are directly linked to the production activity; they may differ according to local conditions.

2. On the basis of these costs, one can estimate a "break-even cost", or "supply cost", the market value required to provide an adequate real return on capital for a new project.

3. Conventional and unconventional gases are expected to be in the same range, as conventional resources are depleted and new projects are moving to the more expensive Norwegian Arctic.

4. This estimate for China applies to both conventional and unconventional gas, as the easy conventional gas is depleting and production is moving to off-shore or more remote regions.

5. The average EUR varies from one shale to another, but also depends on the experience of the industry in a given shale; with time, the industry optimises the technologies used and extracts more gas from each well.

**Ⅴ. Put the following paragraphs into Chinese.**

1. Capital costs, often called finding and development costs, are usually dominated by the costs of constructing wells. Shale gas wells do cost more than conventional gas wells in the same conditions, because of the additional costs of multistage hydraulic fracturing; the same consideration applies to tight gas wells, for the same reason. Coalbed methane wells have so far been relatively cheap, compared with conventional gas wells, because production has been at shallow depths in regions with well-developed markets.

2. Import infrastructure in excess of 100 bcm was built in the United States alone in this period, with matching LNG supply investments in major producers, such as Qatar. However, in 2011, net LNG imports to North America were less than 20 bcm, out of a total market exceeding 850 bcm; 8 bcm into the United

States and 9 bcm into Mexico and Canada. Hence, major quantities of LNG supply became available for other global markets, including Asia and Europe.

## 6.2 Mineral Licensing and Leasing

### Guidance to Reading

*Accessing mineral rights is a key step in the shale gas development process. Without this access, gas cannot be legally extracted and marketed for sale. Differences in mineral rights ownership around the world lend regional shape to the oil and gas industry, have enormous economic affects within countries, and have contributed to the ability of the United States to take the early lead in the development of processes and technologies for the exploration and extraction of shale gas. The difference in mineral rights ownership between the United States and the rest of the world is a key factor to make the US shale gas industry lead the world.*

### Text

**Mineral leases** are contractual agreements between the owner of a mineral estate and oil and gas companies to provide the companies the right to explore for, develop, and extract from mineral deposits in the area described in the **lease**. Mineral rights can be either owned by governments or by private entities including businesses and private citizens. In fact, private individuals, rather than government, own much of the mineral rights across the country. Additionally, these mineral rights can be separated from surface rights(with agreement from the mineral rights owner), meaning that mineral rights can be sold **outright** or leased.

#### *Government-owned Mineral Rights*

The specific procedures for the management of access to government-owned mineral rights for oil and gas development vary from country to country. In general, governments section their lands into **discrete** parcels and then offer the licenses, subject to various terms and conditions, to oil and gas companies to allow the exploration for and eventual extraction of oil and gas found underground. Typically, license terms and conditions include **commitment**, sometimes in the form of a specified investment amount, by the company to conduct certain exploration activities. In exchange, the company obtains a multiple year right to conduct exploration activities in an attempt to discover commercially viable quantities of oil or gas. If exploration activities are successful, the company then **petitions** the government for permission to develop the license. This petition usually involves a detailed development plan including environmental impact statements and planning for wells, plants, and infrastructure. The government will negotiate an interest in the gas field, and convert the exploration license into a multi-year( often 15 to 20 years) production license. Owing in part to the bureaucratic nature of this process and varied circumstances around the world, the process for **obtaining access to** governmentowned mineral rights **is fraught with** opportunity for delay and can require years to complete.

### *Privately Owned Mineral Rights*

Mineral leasing with private owners typically involves lease payments to the owner of the mineral estate at the time of signing to reserve the property for the **lessee** for a determined time period. In the event that commercially viable quantities of gas are found to exist in the leased property, the company may proceed to develop the property for production. Royalties are generally included in mineral leases in which the owner is paid a share of the wellhead value of gas production. If production is not **commenced prior to lease expiration**, all rights to the property and minerals **revert** to the owner unless the lease is renewed.

One-time lease payments are typically made on a per-acre basis and royalties assessed as a percentage of wellhead value of gas production, assuming wells are eventually drilled and produce gas. Two key factors affecting lease payment and royalty amounts are the lessee's level of confidence that a shale will produce and demand for mineral leases. In the early stages of development of a shale play, lessees tend to offer more **modest** payments as they **edge** against uncertainty that a specific area will become commercially **viable**. As the play becomes increasingly proven and developed, lease and royalty payments rise with demand from gas companies for access to mineral rights. Another factor affecting royalty amounts is the existence of regulation governing leases in a particular geography.

### *Split Estate*

In the cases of both government and privately owned mineral rights, the owner of the mineral rights on a specific property may be different than the owner of the surface rights, or land that lies above the minerals, which creates a "**split estate.**" The condition of split estate exists in most countries of the world apart from the United States. Within the US, the **commonness** of government-private citizen split estate varies by region. For instance, the federal government owns much of the mineral rights in western states, but only 4 percent of the lands in the eastern and midwestern states, where US shale gas resources are most prevalent, are associated with a federal split estate. Split-estate circumstances between private owners can, however, occur as mineral rights and surface rights can be transferred independently in the US.

Conflict can arise in split estate situations as gas development can require use of the surface land above a mineral estate. In such cases, the mineral lease holder **is** generally **entitled to** reasonable use of the land surface although they do not own the surface property.

### Mineral Rights Access Has Shaped theUS Shale Gas Industry

The difference in mineral rights ownership between the United States and the rest of the world is one factor which makes the shale gas industry in the US differ from that of other regions around the world. The ownership structure allows entities to approach any mineral rights owner at virtually any time to negotiate an agreement, rather than only when oil and gas exploration licenses are issued by the government. Further, the profit incentive associated with the production of gas can create a pool of motivated mineral estate owners, resulting in a substantially shorter leasing process. In the United States, obtaining mineral leases typically requires months instead of years as in other countries. This simultaneously improves shale gas economics, reduces barriers to entry to the gas industry and creates opportunities for small to medium sized independent gas producers.

Since royalties account for the most significant upside in these deals, companies that are able to convince owners of mineral estates that they are most likely to produce quickly have an advantage in obtaining early access. A deal with a company which appears more efficient and motivated to develop shale resources, such as an independent E&P company with relatively limited resource diversity in its portfolio, can be more attractive to a mineral estate owner than one with a multi-billion dollar **supermajor** that may or may not **prioritize** development in the near term. This is another factor that has led to the proliferation of small to medium sized independent gas companies in the US shale plays over the supermajors to date.

### Well Permitting

Once the right to **exploit** a shale gas resource has been secured, developers seek permits to **authorize** drilling of wells. In the United States, these permits are issued by the individual states in which the well is to be drilled. State regulatory agencies have the authority to regulate and permit shale gas development activities in accordance with federal and state laws. State regulators permit well design, location, spacing, operation, and abandonment and permit applications require specific information with regard to each of these elements be detailed. These agencies also issue permits related to water management, waste management and disposal, air emissions, underground injection, wildlife impacts, surface disturbance, and worker health and safety. Finally, other levels of government, such as cities, counties, or regional water authorities, may require permits in addition to those granted by the state regulators. Such permits typically address issues such as the placement of wells in floodzones and set backs from residences.

## Words and Expressions

| | |
|---|---|
| lease | 租约;租契 |
| outright | 率直的;完全地;彻底地 |
| discrete | 分离的,不相关联的 |
| commitment | 承担义务;承诺 |
| petition | 祈求,请求,请愿 |
| lessee | 承租人 |
| commence | 开始 |
| revert | 恢复;复归 |
| modest | 节制的;适度的; |
| edge | 侧身移动;徐徐移动 |
| viable | 有生存力的 |
| commonness | 普通,平凡,共性 |
| supermajor | 石油巨头 |
| prioritize | 把……区分优先次序 |
| exploit | 开发;利用;开拓 |
| authorize | 授权,批准,委托 |

## Phrases and Expressions

| | |
|---|---|
| mineral lease | 矿产租赁协议 |

| obtain access to | 获得……权利 |
| be fraught with | 充满 |
| prior to | 在……之前 |
| lease expiration | 租约期满 |
| split estate | 分割地产 |
| be entitled to | 有……的使用权 |

## Language Focus

1. In general, governments section their lands into discrete parcels and then offer the licenses, subject to various terms and conditions, to oil and gas companies to allow the exploration for and eventual extraction of oil and gas found underground.

（参考译文：一般情况下，政府把其所拥有的土地划分为不连续的区块，然后附带各种不同条款和条件，给油气公司颁发许可，允许其勘探并最终开发地下油气资源。）

本句为简单句，主语 governments 带有两个谓语动词 section 和 offer，构成一个主谓结构。subject to various terms and conditions 为形容词短语做插入语，subject to 是"服从"的意思；to allow 不定式为目的状语，oil and gas 既是 extraction of 的介词宾语，又是 exploration for 的介词宾语，found underground 为过去分词短语做 oil and gas 的后置定语。

2. Owing in part to the bureaucratic nature of this process and varied circumstances around the world, the process for obtaining access to governmentowned mineral rights is fraught with opportunity for delay and can require years to complete.

（参考译文：要获得国有矿产开发权很有可能需要数年时间才能完成，其部分原因是世界各地存在着不同程度的官僚作风及其他情况。）

本句中 Owing in part to the bureaucratic nature of this process and varied circumstances around the world 短语做原因状语，in part 意思是"部分地"；Owing to 意思是"由于……；因为……"。

3. For instance, the federal government owns much of the mineral rights in western states, but only 4 percent of the lands in the eastern and midwestern states, where US shale gas resources are most prevalent, are associated with a federal split estate.

［参考译文：例如，联邦政府拥有西部各州大部分矿产权，但在东部和中

西部各州(美国页岩气资源最多的地区),只有4%的土地涉及联邦政府地产分离。]

本句是由 but 连接的并列句,后面分句中的主句主谓结构是 only 4 percent of the lands... are associated with...,主语后有一个 where 引导的非限制性定语从句,修饰主语 only 4 percent of the lands in the eastern and midwestern states。

4. A deal with a company which appears more efficient and motivated to develop shale resources, such as an independent E&P company with relatively limited resource diversity in its portfolio, can be more attractive to a mineral estate owner than one with a multi-billion dollar supermajor that may or may not prioritize development in the near term.

(参考译文:一个在页岩气开发方面更积极、更高效的公司,如其产品组合中资源多样性相对有限的独立勘探开发公司,比拥有数十亿美元资产,但不一定会在短期内优先开发页岩气的大公司更加吸引矿主与其达成交易。)

本句主句主谓结构为 A deal... can be more attractive,介词短语 with a company 是主语的后置定语,which 引导的从句做 company 的后置定语,such as... 是插入语,表示例子。than 引出比较状语从句,比较分别和两类公司做的交易,one 指代 a deal,that 从句是 supermajor 的后置定语。

5. State regulators permit well design, location, spacing, operation, and abandonment and permit applications require specific information with regard to each of these elements be detailed.

(参考译文:气井的设计、位置、间距、操作和遗弃都要由各州监管机构批准,审批过程中需要提供有关方面的详细信息。)

本句为并列句,由 and 连接。后面分句中的主语是 permit applications,谓语是 require,宾语是虚拟结构的从句,引导词 that 省略了,be detailed 是从句谓语,be 前面省略了 should。

## Reinforced Learning

### I. Answer the following questions for a comprehension of the text.

1. Describe mineral leases according to the text.

2. Describe the so called "Split Estate".

3. Why can conflict arise in split estate situations?

4. Please list at least one factor that has led to the proliferation of small to medium sized independent gas companies in the US shale plays over superma-

jors to date.

5. Who issues the permits to authorize drilling of wells?

**Ⅱ. Multiple choice: choose the correct one from the alternative answers to give the exact meaning of the words.**

1. Owing in part to the bureaucratic nature of this process and varied circumstances around the world, the process for obtaining access to governmen-towned mineral rights is <u>fraught with</u> opportunity for delay and can require years to complete.

  A. full of   B. fractured with C. crowded with  D. brought up

2. If production is not commenced prior to lease <u>expiration</u>, all rights to the property and minerals revert to the owner unless the lease is renewed.

  A. experience B. explanation  C. termination  D. expression

3. The ownership structure allows entities to approach any mineral rights owner at <u>virtually</u> any time to negotiate an agreement, rather than only when oil and gas exploration licenses are issued by the government.

  A. completely B. literally   C. vigorously   D. surely

4. Further, the profit <u>incentive</u> associated with the production of gas can create a pool of motivated mineral estate owners, resulting in a substantially shorter leasing process.

  A. increase  B. amazing   C. inducement  D. basic

5. This simultaneously improves shale gas economics, reduces <u>barriers</u> to entry to the gas industry and creates opportunities for small to medium sized independent gas producers.

  A. foundation B. battery   C. discussion   D. obstacles

6. Once the right to <u>exploit</u> a shale gas resource has been secured, developers seek permits to authorize drilling of wells.

  A. explain  B. tap    C. excuse    D. explore

7. These agencies also <u>issue</u> permits related to water management, waste management and disposal, air emissions, underground injection, wildlife impacts, surface disturbance, and worker health and safety.

  A. account  B. release   C. isolate    D. proceed

8. Finally, other levels of government, such as cities, counties, or regional water authorities, may require permits in addition to those <u>granted</u> by the state regulators.

A. appraised　　B. permitted　　C. announced　　D. published

9. Such permits typically address issues such as the placement of wells in floodzones, set backs from <u>residences</u> or other protected sites, noise level, site maintenance, and traffic.

A. work　　B. housings　　C. place　　D. country

10. For instance, the federal government owns much of the mineral rights in western states, but only 4 percent of the lands in the eastern and midwestern states, whereUS shale gas resources are most <u>prevalent</u>, are associated with a federal split estate.

A. substantial　　B. functional　　C. pregnant　　D. common

**Ⅲ. Multiple choice: read the four suggested translations and choose the best answer.**

1. Mineral leases are <u>contractual</u> agreements between the owner of a mineral estate and oil and gas companies to provide the companies the right to explore for, develop, and extract from mineral deposits in the area described in the lease.

A. 合同的　　B. 系统的　　C. 口头的　　D. 书面的

2. Owing in part to the <u>bureaucratic</u> nature of this process and varied circumstances around the world, the process for obtaining access to governmen-towned mineral rights is fraught with opportunity for delay and can require years to complete.

A. 虚幻的　　B. 现实的　　C. 资本的　　D. 官僚的

3. This <u>simultaneously</u> improves shale gas economics, reduces barriers to entry to the gas industry and creates opportunities for small to medium sized independent gas producers.

A. 异步地　　B. 同时地　　C. 先后地　　D. 缓慢地

4. A deal with a company which appears more efficient and motivated to develop shale resources, such an independent E&P company with relatively limited resource diversity in its <u>portfolio</u>, can be more attractive to a mineral estate owner than one with a multi-billion dollar supermajor that may or may not prioritize development in the near term.

A. 计划　　B. 投资组合　　C. 目标　　D. 资产负债

5. In general, governments section their lands into discrete <u>parcels</u>.

A. 监管　　B. 地块　　C. 薄片　　D. 权利

## Ⅳ. Put the following sentences into Chinese.

1. Mineral leases are contractual agreements between the owner of a mineral estate and oil and gas companies to provide the companies the right to explore for, develop, and extract from mineral deposits in the area described in the lease.

2. Additionally, these mineral rights can be separated from surface rights (with agreement from the mineral rights owner), meaning that mineral rights can be soldoutright or leased.

3. The specific procedures for the management of access to government-owned mineral rights for oil and gas development vary from country to country.

4. Conflict can arise in split estate situations as gas development can require use of the surface land above a mineral estate.

5. State regulators permit well design, location, spacing, operation, and abandonment and permit applications require specific information with regard to each of these elements be detailed.

## Ⅴ. Put the following paragraphs into Chinese.

1. In the United States, the profit incentive associated with the production of gas can create a pool of motivated mineral estate owners, resulting in a substantially shorter leasing process. Obtaining mineral leases typically requires months instead of years as in other countries. This simultaneously improves shale gas economics, reduces barriers to entry to the gas industry and creates opportunities for small to medium sized independent gas producers.

2. In the cases of both government and privately owned mineral rights, the owner of the mineral rights on a specific property may be different than the owner of the surface rights, or land that lies above the minerals, which creates a "split estate." The condition of split estate exists in most countries of the world apart from the United States. Within the US, the commonness of government-private citizen split estate varies by region. For instance, the federal government owns much of the mineral rights in western states, but only 4 percent of the lands in the eastern and midwestern states, where US shale gas resources are most prevalent, are associated with a federal split estate. Split-estate circumstances between private owners can, however, occur as mineral rights and surface rights can be transferred independently in the US.

# 6. 3　Market Demand Drivers

## Guidance to Reading

*Being more efficient and less carbon-intensive than coal, natural gas is becoming more important as a fuel for electricity generation. Driven by recent growing supplies of shale gas, natural gas prices have decreased, further promoting coal-to-gas fuel switching. As world oil prices are expected to continue to rise for the next several decades, less expensive natural gas is expected to substitute oil where possible. Even if only 25 percent of the estimated total resource is technically recoverable, shale gas alone can meet global natural gas demand for over 40 years at current consumption levels. Although conventional oil and gas resources are* **unevenly** *distributed around the world, the widely distributed shale resources offer energy importing nations the potential to increase their domestic production, and enhance overall energy security.*

## Text

### Growth in Global Energy Consumption

Global energy consumption is expected to increase by 25% from 473 **quadrillion** Btu in 2005 to over 590 quadrillion Btu in 2020. Global recessionary conditions moderated consumer demand for goods and services and related manufacturing, and global energy consumption **concurrently** declined by 2.2% in 2009. Since then, the developing non OECD ( **Organization for Economic Cooperation and Development**) nations have led the global recovery. There are also indications that the **recession** in the US has ended, and while recovery in Europe and Japan has lagged, the EIA International Energy Outlook projects that most nations will have recovered from recessionary conditions by 2012, and revert to long-term growth **patterns**.

OECD countries, with developed and mature energy infrastructures, have historically accounted for the largest share of world energy consumption. However, the growth of energy consumption in the non-OECD economies has **outpaced** that of the OECD for several years. China and India are projected to lead energy demand in the non-OECD nations and have the highest growth rates. Substantial growth is also expected in non-OECD nations in the Middle East and Central and

South America. By 2020, non-OECD economies will account for nearly 57 percent of world energy consumption.

### Growth in Global Natural Gas Demand

Natural gas is expected to grow in importance as a fuel for electricity generation worldwide as it is both more efficient and less carbon-intensive than coal. The **substitution** of coal-fired power plants with natural gas plants has been ongoing in the US over the past decade and continues **unabated**. Over the recent past, the increase in delivered coal prices and the decrease in delivered natural gas prices, driven largely by growing supplies of shale gas, further aided coal-to-gas fuel switching. In the US, coal-fired electric power generation declined 11.6 percent from 2008 to 2009, bringing coal's share of the electricity power output to 44.5 percent, the lowest level since 1978. **In contrast**, natural gas fueled generation rose 4.3 percent in 2009, despite the 4.1 percent decline in overall electric generation, increasing the natural gas share of generation to 23.3 percent, the highest level since 1970. Electricity's share of total US natural gas consumption has also risen rapidly, growing from 17 percent in 1996 to over 30 percent in 2009. As shale gas production increases the availability of natural gas, such coal-to-gas substitution is expected to occur in many nations around the world.

In addition to its use as a fuel source in electricity generation, natural gas is commonly used as a fuel for residential, industrial and commercial heating and cooling, as a **feedstock** for various **petrochemical** products, and as an alternative fuel for the transportation sector. As world oil prices are expected to continue to rise for the next several decades, it is expected that less expensive natural gas will substitute oil where possible. Newly constructed petrochemical plants in the industrial sector are expected to rely increasingly on natural gas as a feedstock. Further, the development and adoption of alternate fuel vehicles that leverage **compressed natural gas** may also significantly increase the demand for natural gas.

The amount of energy sourced from natural gas is expected to further increase by 22 percent as global demand for natural gas continues to grow by 2 percent per year between 2010 and 2020. At over 2.8 percent per year, consumption growth in the non-OECD economies is expected to be nearly 3 times faster than that in OECD nations, which is projected to grow at 1.1 percent per year.

### Decline in OECD Conventional Natural Gas Supplies

While demand and consumption of natural gas continues to grow, the supply and production of natural gas from conventional resources are declining, particularly in the developed OECD nations of Europe and North America. EIA projections anticipate that the production of conventional natural gas in OECD nations will decline by 11 percent between 2008 and 2020. While substantial growth in conventional gas production is anticipated in developing non-OECD nations, much of the production growth is expected to be domestically consumed and not available for exports. Further, even in cases where surplus production is available, the means of transporting natural gas from nations with excess production to those with short supplies are often unavailable, uneconomical, or fraught with geo-political constraints. Thus, while natural gas is actively traded between Canada, the US and Mexico in the North American region, pipeline natural gas trade between North America and the Middle East is non-existent. Similarly, significant Russian natural gas production is piped into Europe, but Japan is unable to import natural gas through pipelines.

### *Advances in Shale Gas Technologies*

While the existence of substantial shale gas resources in various regions of the world has been known for decades, the technological and economic **feasibility** of recovering commercial quantities of shale gas from these deep, low-permeability, unconventional resources had limited development. Processes and technologies specific to shale gas exploration, drilling, resource stimulation, extraction, and environmental impact management developed and successfully deployed in the US over the past decade have demonstrated the viability of shale gas production. Several innovative technologies have been **implemented** in shale gas development with advanced exploration, drilling, and well stimulation technologies widely regarded as the most influential. The application of technologies including 3D and 4D seismic exploration, directional drilling, and multi-stage hydraulic fracturing have resulted in a dramatic increase in the US production of shale gas over the past decade.

### *Abundance of Global Shale Gas Resources*

The total global shale gas resource base is estimated to exceed 18,775 trillion cubic feet(Tcf), as shown in Frgure 6. 1. This represents the total amount of natural gas estimated to exist in shale formations around the world and includes both currently discovered and as yet undiscovered shale resources. Even if only 25 percent of the estimated total resource base becomes technically recoverable, shale gas alone can meet global natural gas demand for over 40 years at current consumption levels.

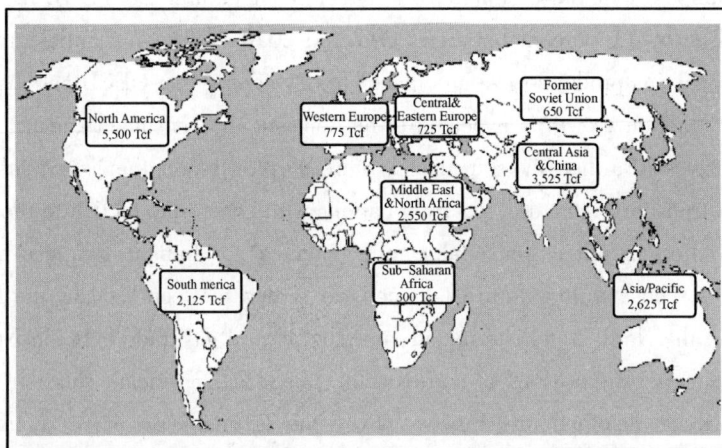

Figure 6.1    Global Distribution of Shale Gas Resources

Conventional oil and gas resources that are unevenly distributed around the world, with the majority of resources concentrated in a handful of countries and resulting imbalances have caused ongoing geo-political **strife**. Widely distributed shale resources offer several energy importing nations the potential to increase domestic production, reduce dependence on imports, and enhance overall energy security. It is now more **cost – efficient** to process and ship LNG than it is to pipe natural gas across great distances. The combination of large potential shale gas supplies and growing use of LNG transportation infrastructure is likely to reshape global natural gas markets and energy politics over the coming decades.

## 🔲 Words and Expressions

| | |
|---|---|
| unevenly | 不均匀地 |
| quadrillion | 千的五次方 |
| concurrently | 同时地 |
| recession | 经济衰退,不景气 |
| pattern | 样品;花样;方式 |
| outpaced | 超过……速度,赶过(outpace 的过去式和过去分词) |
| substitution | 代替;代用;替换 |

| unabated | 不衰的,不减弱的 |
| --- | --- |
| feedstock | 给料(指供送入机器或加工厂的原料);进料 |
| petrochemical | 石油化学产品 |
| feasibility | 可行性 |
| implement | 实现;执行;使……生效 |
| strife | 争吵 |

## Phrases and Expressions

| OECD(Organization for Economic Cooperation and Development) | 经济合作与发展组织 |
| --- | --- |
| in contrast(to, with) | 与……形成对比,与……对比起来 |
| compressed natural gas(CNG) | 压缩天然气 |
| cost-efficient(cost-effective) | 有成本效益的;值得花钱的;划算的 |

## Language Focus

1. There are also indications that the recession in the US has ended, and while recovery in Europe and Japan has lagged, the EIA International Energy Outlook projects that most nations will have recovered from recessionary conditions by 2012, and revert to long-term growth patterns.

(参考译文:还有迹象表明,美国经济衰退已经结束,但因欧洲和日本经济复苏滞后,EIA 国际能源展望预测:到 2012 年世界大部分国家会从经济衰退中恢复,并重新回到长期增长模式。)

本句为 and 连接的并列复合句,前面分句为倒装结构,主语是 indications,谓语是 There are,that 从句为主语的后置定语。后面分句是复合句,主句是 the EIA International Energy Outlook projects that..., 其中包含一个 that 宾语从句,该从句主语带有两个谓语,will have recovered 和(will)revert to;该分句还有一个 while 引导的让步状语从句。

2. In addition to its use as a fuel source in electricity generation, natural gas is commonly used as a fuel for residential, industrial and commercial heating and cooling, as a feedstock for various petrochemical products, and as an alternative fuel for the transportation sector.

(参考译文:除了用来发电外,天然气也普遍用于居民住宅、工业和商业供暖及制冷,另外还可用作石化产品原料以及运输替代燃料。)

本句中,in addition to 介词短语作状语,意为"除了……之外";as..., as..., and as... 为并列平行结构,说明 natural gas 的用途。

3. While the existence of substantial shale gas resources in various regions of the world has been known for decades, the technological and economic feasibility of recovering commercial quantities of shale gas from these deep, low-permeability, unconventional resources had limited development.

(参考译文:虽然发现世界各地有大量页岩气资源已有数十年了,但在这些埋藏深且渗透率又低的非常规资源中,进行页岩气商业开采的技术和经济上的可行性都很有限。)

本句中,While 引导让步状语从句,意思是"尽管,虽然"。

4. Processes and technologies specific to shale gas exploration, drilling, resource stimulation, extraction, and environmental impact management developed and successfully deployed in the US over the past decade have demonstrated the viability of shale gas production.

(参考译文:美国在过去的十年中成功开发并应用了页岩气勘探、钻井、增产、开采和环境治理等工艺和技术,证明了页岩气生产的可行性。)

本句主谓结构为 Processes and technologies... have demonstrated the viability of shale gas production,形容词短语 specific to shale gas exploration... 做后置定语修饰 processes and technologies,过去分词短语 developed and successfully deployed in the US over the past decade 也是 processes and technologies 的后置定语。

5. It is now more cost efficient to process and ship LNG than it is to pipe natural gas across great distances.

(参考译文:现在将天然气液化后船运比通过管道远距离运输更经济。)

本句含有 than 引导的比较级状语从句,比较的双方是 to process and ship LNG 和 to pipe natural gas across great distances。

## ▣ Reinforced Learning

### Ⅰ. Answer the following questions for a comprehension of the text.

1. Which non-OECD nations are expected to have the highest energy demand and also the highest growth rates?

2. Give at least one example to prove that the means of transporting natural gas from nations with excess production to those with short supplies are often

unavailable, uneconomical, or fraught with geo-political constraints.

3. List various usages of natural gas according to the text.

4. What has limited the development of shale gas recovery?

5. What willthe world unevenly distributed oil and gas resources bring about?

**Ⅱ. Multiple choice: choose the correct one from the alternative answers to give the exact meaning of the words.**

1. Global energy underline consumption is expected to increase by 25% from 473 quadrillion Btu in 2005 to over 590 quadrillion Btu in 2020.

A. save　　　　B. stock　　　　C. expenditure　　D. pardon

2. There are also indications that the recession in the US has ended, and while recovery in Europe and Japan has lagged, the EIA International Energy Outlook projects that most nations will have recovered from recessionary conditions by 2012, and revert to long-term growth patterns.

A. experienced　B. exceeded　　C. taken　　　　D. fallen behind

3. There are also indications that the recession in the US has ended, and while recovery in Europe and Japan has lagged the EIA International Energy Outlook projects that most nations will have recovered from recessionary conditions by 2012, and revert to long-term growth patterns.

A. complete　　B. fall　　　　C. return　　　　D. put

4. The substitution of coal-fired power plants with natural gas plants has been ongoing in the US over the past decade and continues unabated.

A. in progress　B. incredible　C. fast　　　　D. basic

5. EIA projections anticipate that production of conventional natural gas in OECD nations will decline by 11 percent between 2008 and 2020.

A. explain　　B. quarrel　　　C. discuss　　　D. foresee

6. Further, even in cases where surplus production is available, the means of transporting natural gas from nations with excess production to those with short supplies are often unavailable, uneconomical, or fraught with geo-political constraints.

A. enough　　B. more　　　　C. extra　　　　D. various

7. Several innovative technologies have been implemented in shale gas development with advanced exploration, drilling, and well stimulation technologies widely regarded as the most influential.

A. groundbreaking　　　　　　B. efficient

  C. isolated        D. economic

  8. Conventional oil and gas resources that are unevenly distributed around the world, with the majority of resources concentrated in a <u>handful</u> of countries and resulting imbalances have caused ongoing geo-political strife.

  A. smattering    B. lot      C. much     D. quantity

  9. In contrast, natural gas fueled generation rose 4.3 percent in 2009, <u>despite</u> the 4.1 percent decline in overall electric generation, increasing the natural gas share of generation to 23.3 percent, the highest level since 1970.

  A. seeing    B. because of   C. owing to     D. in spite of

  10. As shale gas production increases the availability of natural gas, such coal-to-gas <u>substitution</u> is expected to occur in many nations around the world.

  A. substance    B. replacement   C. pregnancy    D. awkward

**III. Multiple choice : read the four suggested translations and choose the best answer.**

  1. Global <u>recessionary</u> conditions moderated consumer demand for goods and services and related manufacturing, and global energy consumption concurrently declined by 2.2% in 2009.

  A. 有序的     B. 衰退的    C. 繁荣的     D. 奢华的

  2. OECD countries, with developed and mature energy infrastructures, have <u>historically</u> accounted for the largest share of world energy consumption.

  A. 在本质上   B. 在历史上    C. 在经济上    D. 在制度上

  3. The substitution of coal-fired power plants with natural gas plants has been ongoing in the US over the past decade and continues <u>unabated</u>.

  A. 不尽力的   B. 不落后的    C. 不减退的    D. 不轻易的

  4. Processes and technologies specific to shale gas exploration, drilling, resource stimulation, extraction, and environmental impact management developed and successfully deployed in the US over the past decade have demonstrated the <u>viability</u> of shale gas production.

  A. 计划     B. 效率      C. 生存能力   D. 切实可行

  5. Widely distributed shale resources offer several energy importing nations the potential to increase <u>domestic</u> production, reduce dependence on imports, and enhance overall energy security.

  A. 国外的     B. 国内的     C. 境外的     D. 全球的

## Ⅳ. Put the following sentences into Chinese.

1. Over the recent past, the increase in delivered coal prices and the decrease in delivered natural gas prices, driven largely by growing supplies of shale gas, further aided coal-to-gas fuel switching.

2. As world oil prices are expected to continue to rise for the next several decades, it is expected that less expensive natural gas will substitute oil where possible.

3. While substantial growth in conventional gas production is anticipated in developing non-OECD nations, much of the production growth is expected to be domestically consumed and not available for exports.

4. While the existence of substantial shale gas resources in various regions of the world has been known for decades, the technological and economic feasibility of recovering commercial quantities of shale gas from these deep, low-permeability, unconventional resources had limited development.

5. Conventional oil and gas resources that are unevenly distributed around the world, with the majority of resources concentrated in a handful of countries and resulting imbalances have caused ongoing geo-political strife.

## Ⅴ. Put the following paragraphs into Chinese.

1. Several innovative technologies have been implemented in shale gas development with advanced exploration, drilling, and well stimulation technologies widely regarded as the most influential. The application of technologies including 3D and 4D seismic exploration, directional drilling, and multi-stage hydraulic fracturing have resulted in a dramatic increase in the US production of shale gas over the past decade.

2. Conventional oil and gas resources that are unevenly distributed around the world, with the majority of resources concentrated in a handful of countries and resulting imbalances have caused ongoing geo-political strife. Widely distributed shale resources offer several energy importing nations the potential to increase domestic production, reduce dependence on imports, and enhance overall energy security. The combination of large potential shale gas supplies and growing use of LNG transportation infrastructure is likely to reshape global natural gas markets and energy politics over the coming decades.

# 6.4　Market Demand Inhibitors

## Guidance to Reading

*The development of natural gas from shale has proceeded quickly in the United States. However, the exploration and extraction of shale gas resources remains a relatively new and developing field. The exploration and development of these resources outside of the United States are still in the early stages. Certain challenges such as high development costs remain to be overcome and debate continues on the estimates of the scale of recoverable shale resources, the economics of production, the environmental impact of shale gas extraction processes and the resulting regulatory responses. All these factors will inhibit the market demand for shale gas. Moreover, adverse policies regarding land and mineral development could also impede growth.*

## Text

### Low Natural Gas Prices Can Impede Shale Growth

Shale well drilling and completion has historically been more expensive, with drilling and completion costs of $ 2 to $ 3 million per well. In order to make reasonable profits, operators need to be able to sell their production at relatively high prices. It is important to note that as industry operators continue to gain experience, and as new technologies continue to improve production rates and lower costs, the break-even price associated with profitable shale production has the potential to decrease further.

If natural gas prices remain low, reduced expectations of shale investments may limit capital **expenditures** and, ultimately, the rate of growth in shale gas production. While specific shale development costs and natural gas prices vary across regions of the world, the **underlying** principles of supply side economics remain similar, and have the potential to impact global shale gas investments and growth and represent a material risk factor.

### Environmental Issues and Regulatory Response

Sustainable production of shale gas from deep, low-permeability resources presents significant environmental challenges. Many of these environmental impacts to land, water, and air are **inherent** to the processes commonly used in oil and gas exploration and production, and yet others result from the specific technologies deployed in the shale gas resource development process. Environmental issues represent a risk factor to the future growth of the shale gas industry, particularly if these issues result in increasingly restrictive regulation.

In the United States, several key federal regulations govern various environmental aspects of oil and gas development. These include:

- **Clean Air Act( CAA )** , which limits air emissions from engines, gas processing equipment, and other sources associated with drilling and gas production.

- **Clean Water Act( CWA )** , which regulates surface **discharges** of water associated with construction sites, as well as storm water **runoff** from production sites. However, the Energy Policy Act of 2005 **exempted** the oil and gas well pad construction and drilling activities **from** federal stormwater permit regulations and such permitting is now addressed variously at the state level.

- **Safe Drinking Water Act( SDWA )** , which generally regulates the underground injection of fluids from oil and gas activities. However, the Energy Policy Act of 2005 exempted hydraulic fracturing fluids, except where they contain **diesel** fuel, from federal regulation under the safe water drinking act.

- **National Environmental Policy Act( NEPA )** , which requires exploration and production on federal lands to be thoroughly analyzed for environmental impacts.

As gas development has moved toward more heavily populated areas with the increased production of gas from shale, environmental activists have **enlisted** the support of concerned local community members to increase awareness of the risks associated with technologies specific to shale gas development. Many of these activists support increased regulatory action, maintaining that localities do not adequately regulate the potential effects of shale gas development. The debate largely **centers around** the water impacts of shale development, but air quality and land surface disturbances are also receiving attention. However, shale industry proponents maintain that the current **regime** of regulatory **oversight** is adequate and additional requirements would negatively affect the economics of shale gas production.

### *Evolving Regulatory Response*

Various shale gas producing states have taken regulatory action **in response to** growing environmental concerns and regulatory activity has been **accelerating**. Actions range from proposal of regulations that would affect new equipment or changes that would increase emissions to establishing **moratoriums** on new shale gas drilling until environmental impact studies are concluded. **Compliance with** these regulations would be required in order for gas companies to obtain drilling permits. Some examples include:

- On July 29, 2010, the Texas Commission on Environmental Quality proposed new rules aimed at ensuring that emissions from shale gas drilling in the Barnett do not **adversely** impact air quality.

- On August 16, 2010, the governor of Wyoming approved new rules covering a variety of drilling practices, including hydraulic fracturing, requiring operators to provide details of the chemical additives used in the fracturing process.

- In October 2010, Michigan Department of Natural Resources and Environment imposed special permitting requirements for wells that are to be hydraulically fractured.

- **As of** November 2010, the New York Assembly is **on the verge of** imposing a moratorium on new permits for natural gas wells that use hydraulic fracturing.

Nonetheless, the gas industry **is wary of** potential additional federal regulation, which is viewed by the industry as unnecessary given the regulations imposed by the states. The industry is further concerned that the costs of compliance with federal regulation would adversely affect the economics of shale gas production. Research conducted by the Independent Petroleum Association of America suggests that federal regulation could result in the addition of about $ 100,000 in cost to each new natural gas well. Environmental issues represent a risk factor to the future growth of the shale gas industry, particularly if these issues result in increasingly restrictive regulation.

### *Barriers to Entry*

While shale gas formations are widely distributed around the world, not all shale resources are alike. Geological differences from one resource to another, and often even within different parts of the same resource, can impact the technical feasibility and economic viability of shale development and production.

Critical success factors such as depth and thickness of formation, total organic content, and porosity and permeability can vary widely. These differences **dictate** the feasibility of current technologies, the costs of drilling and fracturing, and well production rates, decline rates, and ultimately the quality and quantity of recoverable gas. Since detailed exploration and evaluation of most shale resources outside North America **have yet to** be conducted, it is likely that a portion of the resources in countries that are in early stages of exploration and evaluation may not be economically or technically recoverable at current time.

Further, since most of the technology development and experiential learning **to date** has occurred in the US, many nations desiring to develop shale gas resources currently do not possess a workforce with shale specific skills and experience. Additionally, the success of shale resource development outside the US may be limited by several other **indigenous** factors such as limitations of current gas processing, storage, and transportation infrastructure, and availability of specialized drilling and fracturing rigs.

Adverse policiesregarding land and mineral development could also impede growth. The specific procedures for the management of access to government-owned mineral rights for gas development vary from country to country. In general, governments section their lands into discrete parcels. Governments then offer these licenses, subject to various terms and conditions, to gas companies to allow the exploration for and eventual extraction of gas found underground. Some countries reserve or **prioritize** access to mineral rights to their own national mining companies, as is traditionally seen in China. Others may only allow access for specific gas development activities and current mineral licensing schemes may not yet address shale development, such as is the case in India.

## 🔲 Words and Expressions

| | |
|---|---|
| expenditure | 支出 |
| underlying | 根本的,基本的;在……下面的 |
| inherent | 固有的 |
| discharge | 排放,流出 |
| runoff | 径流 |
| enlist | 获取(帮助、支持) |
| regime | 管理制度,体制 |
| oversight | 监督 |
| accelerate | 加速 |
| moratorium | 暂停,终止 |
| adversely | 不利地 |
| dictate | 影响,决定 |

indigenous                                       本国的

prioritize                                      优先安排或考虑

## ⊡ Phrases and Expressions

exempt... from                         豁免,免除

center around                           围绕

in response to                        应对

comply with                            服从

as of                                       自……开始;截至

on the verge of                      处于……的边缘,几乎

be wary of                             对……警惕

have yet to do sth                  还没……;尚有待于

to date                                 目前,迄今为止

## ⊡ Language Focus

1. Many of these environmental impacts to land, water, and air are inherent to the processes commonly used in oil and gas exploration and production, and yet others result from the specific technologies deployed in the shale gas resource development process.

（参考译文:许多对土地、水和空气等的环境影响都根源于油气勘探和生产中常用的工艺,还有一些影响是页岩气资源开发过程中使用的特殊技术导致的。）

本句是一个由 and 连接的并列句,两个分句在句末各有一个后置定语。commonly used in oil and gas exploration and production 为过去分词短语作后置定语,修饰 the processes, deployed in the shale gas resource development process 也为过去分词短语作后置定语,修饰 the specific technologies。

2. Actions range from proposal of regulations that would affect new equipment or changes that would increase emissions to establishing moratoriums on new shale gas drilling until environmental impact studies are concluded.

（参考译文:具体行动包括提出监管法规,约束废气排放量增加的新设备或改良设备,以及待环境影响研究结束后再启动新的页岩气钻探项目。）

本句的主句是 Actions **range from** proposal of regulations... **to** establishing moratoriums on new shale gas drilling until environmental impact studies are concluded,说明行动的范围;**that** would affect new equipment or changes **that** would increase emissions 为后置定语从句修饰 regulations,该从句中还有一个定语从句 that would increase emissions,修饰其先行词 new equipment or changes。

3. Nonetheless,the gas industry is wary of potential additional federal regulation,which is viewed by the industry as unnecessary given the regulations imposed by the states.

（参考译文:不过天然气行业反对再有联邦监管法规出台,认为其没有必要,因为各州都有自己的监管法规。）

本句 which is... 为非限制性定语从句,修饰 potential additional federal regulation。given... 表示条件,意为"鉴于"。

## Reinforced Learning

**Ⅰ. Answer the following questions for a comprehension of the text.**

1. Why do natural gas prices affect shale growth?

2. How does the production of shale gas affect the environment?

3. What are the different attitudes of environmental activists and the supporters towards the shale industry?

4. What are the key factors affecting the quality and quantity of recoverable gas?

5. List the barriers to the growth of the shale industry in some countries.

**Ⅱ. Multiple choice:choose the correct one from the alternative answers to give the exact meaning of the words.**

1. If natural gas prices remain low,reduced expectations of shale investments may limit capital expenditures and,underlined{ultimately},the rate of growth in shale gas production.

A. interestingly　　B. finally　　C. surprisingly　　D. expectingly

2. While specific shale development costs and natural gas prices vary across regions of the world,the underlined{underlying} principles of supply side economics remain similar,and have the potential to impact global shale gas investments and growth and represent a material risk factor.

A. basic　　　　B. true　　　　C. genuine　　　　D. firm

3. As gas development has moved toward more heavily populated areas with the increased production of gas from shale,environmental activists have

enlisted the support of concerned local community members to increase awareness of the risks associated with technologies specific to shale gas development.

    A. maintained     B. listed       C. winned       D. obtained

4. The debate largely centers around the water impacts of shale development, but air quality and land surface disturbances are also receiving attention.

    A. controversially  B. mainly     C. constantly     D. continually

5. On July 29, 2010, the Texas Commission on Environmental Quality proposed new rules aimed at ensuring that emissions from shale gas drilling in the
Barnett do not adversely impact air quality.

    A. harmfully     B. contrarily   C. completely    D. further

6. Nonetheless, the gas industry is wary of potential additional federal regulation, which is viewed by the industry as unnecessary given the regulations imposed by the states.

    A. weary       B. afraid      C. cautious     D. aware

7. Compliance with these regulations would be required in order for gas
companies to obtain drilling permits.

    A. making      B. obeying   C. using      D. deploying

8. These differences dictate the feasibility of current technologies.

    A. prescribe    B. influence   C. demand    D. limit

9. Adverse policies regarding land and mineral development could also impede growth.

    A. of        B. about     C. on       D. for

10. In general, governments section their lands into discrete parcels.

    A. divide      B. part      C. depart     D. classify

**III. Multiple choice: read the four suggested translations and choose the
best answer.**

1. Many of these environmental impacts to land, water, and air are inherent
to the processes commonly used in oil and gas exploration and production, and
yet others result from the specific technologies deployed in the shale gas resource development process.

    A. 相关的      B. 固有的   C. 延续的    D. 联系的

2. However, shale industry proponents maintain that the current regime of
regulatory oversight is adequate and additional requirements would negatively
affect the economics of shale gas production.

A. 忽视          B. 调节          C. 管理          D. 监督

3. However, the Energy Policy Act of 2005 <u>exempted</u> the oil and gas well pad construction and drilling activities from federal stormwater permit regulations and such permitting is now addressed variously at the state level.

A. 禁止          B. 废除          C. 豁免          D. 允许

4. Various shale gas producing states have taken regulatory action in response to growing environmental concerns and regulatory activity has been <u>accelerating</u>.

A. 提高          B. 加速          C. 增加          D. 强化

5. Some countries reserve or <u>prioritize</u> access to mineral rights to their own national mining companies, as is traditionally seen in China.

A. 提供          B. 提前          C. 优先考虑          D. 积极获取

**IV. Put the following sentences into Chinese.**

1. Sustainable production of shale gas from deep, low-permeability resources presents significant environmental challenges.

2. Various shale gas producing states have taken regulatory action in response to growing environmental concerns and regulatory activity has been accelerating.

3. Adverse policies regarding land and mineral development could also impede growth.

4. Since detailed exploration and evaluation of most shale resources outside North America have yet to be conducted, it is likely that a portion of the resources in countries that are in early stages of exploration and evaluation may not be economically or technically recoverable at current time.

5. Governments then offer these licenses, subject to various terms and conditions, to gas companies to allow the exploration for and eventual extraction of gas found underground.

**V. Put the following paragraphs into Chinese.**

1. Shale well drilling and completion has historically been more expensive. In order to make reasonable profits, operators need to be able to sell their production at relatively high prices. It is important to note that as industry operators continue to gain experience, and as new technologies continue to improve production rates and lower costs, the break-even price associated with profitable shale production has the potential to decrease further.

2. Many of these activists support increased regulatory action, maintaining that localities do not adequately regulate the potential effects of shale gas development. The debate largely centers around the water impacts of shale development, but air quality and land surface disturbances are also receiving attention. However, shale industry proponents maintain that the current regime of regulatory oversight is adequate and additional requirements would negatively affect the economics of shale gas production.

# Chapter 7　Golden Rules for Unconventional Oil and Gas Development

## 7.1　Drilling Site Construction and Regulation

### 🔲 Guidance to Reading

*Drilling site construction can bring a negative impact to the local ecology, heritage, and existing land use, or may even give rise to earthquakes. Although the risk of triggering an earthquake is small, even minor earth* **tremors** *can easily undermine public confidence in the safety of drilling operations. Unconventional gas production needs the support from society, and the public need to be adequately convinced that the environmental and social risks will be well enough managed. As a result, severe regulations are necessary to ensure that wells are properly designed, constructed and operated. Policy-makers,* **regulators**, *operators and others can follow the golden rules for unconventional drilling site construction to control these environmental and social impacts.*

### 🔲 Text

The outlook for unconventional gas production around the world depends critically on how the environmental issues are addressed. Society needs to be adequately convinced that the environmental and social risks will be well enough managed to warrant consent to unconventional gas production, **in the interests of** the broader economic, social and environmental benefits that the development of unconventional resources can bring.

The Golden Rules below suggest principles that can allow policy-makers, regulators, operators and others to address these environmental and social impacts in order to earn or retain that **consent**.

**Watch where you drill**

*Choose well sites so as to minimise impacts on the local community, heritage, existing land use, individual livelihoods and ecology.* The choice of well site is a moment when engagement with local **stakeholders** and regulators needs to be handled with the utmost care. Each well site needs to be chosen based on the sub-surface geology, but also taking into consideration populated areas, the natural environment and local ecology, existing infrastructure and access roads, water availability and disposal options and seasonal restrictions caused by climate or wildlife concerns. **Sensitivity** at this stage to a range of above-ground concerns can do much to mitigate or avoid problems later in a development.

*Properly survey the geology of the area to make smart decisions about where to drill and where to hydraulically fracture; assess the risk that deep faults or other geological features could generate earthquakes or permit fluids to pass between geological strata.* Careful planning can greatly improve the productivity and recovery rates of wells, reducing the number of wells that need to be drilled and minimising the intensity of hydraulic fracturing and the associated environmental impact. Although the risk of triggering an earthquake is small, even minor earth tremors can easily undermine public confidence in the safety of drilling operations. A careful study of the geology of the area targeted for drilling is necessary to allow operators to avoid operations in areas where deep faults or other characteristics create higher risks.

Producers also need to survey for the presence of oldboreholes or naturally occurring methane in shallow pockets above the source rock and adjust drilling sites(or the pathway of the wellbore) to avoid these areas.

*Monitor to ensure that hydraulic fractures do not extend beyond the gas-producing formations.* The risk of leakage of the fracturing fluid used for shale and tight gas production through the rock from the producing zone into aquifers is minimal because the aquifers are located at much shallower depths; but such migration is theoretically possible in certain exceptional circumstances. A good understanding of the local geology and the use of **micro-seismic**( or other) measuring techniques for monitoring fractures is necessary to minimise the **residual** risk.

*Put in place robust rules on well design, construction, cementing and integrity testing as part of a general performance standard that gas bearing formations must be completely isolated from other strata penetrated by the well, in particular freshwater aquifers.* Regulations need to ensure wells are designed, constructed and operated so as to ensure complete isolation. Multiple measures need to be in place to prevent leaks, with an overarching performance standard requiring operators to follow systematically all recommended industry best practices. This applies up to and including the **abandonment** of the well, i. e. through and beyond the lifetime of the development.

*Consider appropriate minimum-depth limitations on hydraulic fracturing to underpin public confidence that this operation takes place only well away from the water table.* Alongside measures to ensure that wells are designed, built and cemented to a high standard, the regulator may choose to define an appropriate depth limitation for shale and tight gas wells, based on local geology and any risk of communication with freshwater aquifers, above which hydraulic fracturing is prohibited.

Take action to prevent surface spills and leaks from wells, and to ensure that any waste fluids and solids are disposed of properly. This requires both **stringent** regulations and a strong performance **commitment** by all companies involved in drilling and production-related activities to carry out operations to the highest possible standard. Good procedures, training of personnel and ready availability of spill-control equipment are essential to prevent and limit the impact of accidents if they do occur. Upgrading **fluid-disposal systems** so that storage and separation tanks replace open pits (closed-loop systems) can reduce the risk of accidental discharge of wastes during drilling.

Conventional oil and gas developments generally follow a fairly well-defined sequence, but the distinctions between the phases of an unconventional development can be much less clear-cut-development generally proceeds in a more **incremental** fashion. At any given time an operator may be exploring or appraising part of a licence area, developing another part and producing from a third, with different regulatory approvals and permits applying at each stage. The **blurred** lines between the stages of an unconventional resource project development increase the complexity of the interactions between operator and regulators (and between the operator and local communities) throughout the life cycle of the development.

For example, the regulatory system in most **jurisdictions** requires the submission and approval of a detailed field development plan at the end of the exploration phase. However, the longer learning curve for unconventional plays makes it much more difficult to develop **comprehensive** plans at this stage, with the risk that relatively small subsequent alterations might trigger the need to resubmit and reapprove the entire development plan-a lengthy and **burdensome** process for both sides.

Beyond their focus on the proper construction of individual wells and installations, regulators also need to take a broader view of the impact of multiple projects and wells over time. This broader scope is essential when it comes to assessments of water use and disposal and of future water requirements, but can be also required in other areas, including land use, air quality, traffic and noise. In general, a regulatory system that focuses primarily on well-by-well approvals rather than project level authorisations, can fail to provide for some environmental risks and miss opportunities to relieve them. For example, there are investments in infrastructure that may not proceed for an individual well but which would serve appreciably to reduce the cumulative environmental impacts of largescale development, such as centralised water treatment plants or pipeline networks for water supply or removal. One of the ways that a regulatory framework can facilitate this sort of investment is through issuing licences for sufficiently large areas and durations.

Governments are usually instrumental in promoting the **co-ordinated** and timely expansion of regional infrastructure alongside a gas development, ensuring that the regulatory framework serves to encourage or require the construction of gas transportation capacity or an expansion of local power supply. Strong co-ordination and communication is necessary between different branches and levels of government, as the rapid growth of a new industry puts pressure not only on the local physical infrastructure, but also on local social services.

## 🔲 Words and Expressions

| | |
|---|---|
| tremors | 颤抖 |
| regulator | 监管者;校准器;整时器 |
| consent | 准许,赞同;同意;(意见等的)一致 |
| stakeholder | 股东;利益相关者;赌金保管者 |
| sensitivity | 敏感性 |
| micro-seismic | 微震 |
| residual | 残余的;残留的 |
| abandonment | 放弃;抛弃;遗弃;放任 |
| stringent | 严格的;迫切的 |
| commitment | 承诺 |
| incremental | 增加的 |
| blurred | (使)变模糊(blur 的过去式和过去分词) |

| jurisdictions | 裁判权;管辖区域;管辖范围 |
| comprehensive | 广泛的;综合的;有理解力的,悟性好的 |
| burdensome | 繁重的,烦累的,难以承担的;累赘 |
| co-ordinate | 协调 |

## Phrases and Expressions

| in the interests of | 为着……的利益 |
| fluid-disposal system | 泥浆处理系统 |

## Language Focus

1. The outlook for unconventional gas production around the world depends critically on how the environmental issues are addressed.

(参考译文:全球非常规天然气的开发前景主要取决于环境问题如何处理。)

本句中,how 引导宾语从句,作短语 depend on 的宾语。

2. Multiple measures need to be in place to prevent leaks, with an overarching performance standard requiring operators to follow systematically all recommended industry best practices.

(参考译文:要采取多种措施防止泄漏,统一要求作业者系统地遵从推荐的行业标准做法。)

本句中,with 短语是独立主格结构,作状语,standard requiring 在逻辑上为主谓关系。

3. In general, a regulatory system that focuses primarily on well-by-well approvals rather than project level authorisations, can fail to provide for some environmental risks and miss opportunities to relieve them.

(参考译文:一般情况下,监管系统如果侧重于逐井审批,而不是授权项目的话,就可能不会警示某些环境风险,从而错过减小风险的机会。)

本句中 a regulatory system... can fail to... and miss... 是主句主谓结构,that 引导定语从句修饰 system。rather than 相当于 instead of,意思是"不是……(而是)"。

## Reinforced Learning

### Ⅰ. Answer the following questions for a comprehension of the text.

1. To make choice of a well site, which factors should be taken into consideration?

2. What should we do when we survey the geology of the area to decide where to drill and where to hydraulically fracture?

3. What factors may govern the depth limitation for shale and tight gas wells?

4. What could increase the complexity of the interactions between operator and regulators throughout the life cycle of the development?

5. Why is strong co-ordination and communication necessary between different branches and levels of government according to the text?

**II. Multiple choice: choose the correct one from the alternative answers to give the exact meaning of the words.**

1. The outlook for unconventional gas production around the world depends critically on how the environmental issues are addressed.

    A. searchingly     B. seriously     C. disapprovingly   D. crucially

2. Careful planning can greatly improve the productivity and recovery rates of wells, reducing the number of wells that need to be drilled and minimising the intensity of hydraulic fracturing and the associated environmental impact.

    A. supplementary    B. related     C. attendant       D. accompanying

3. Alongside measures to ensure that wells are designed, built and cemented to a high standard, the regulator may choose to define an appropriate depth limitation for shale and tight gas wells, based on local geology and any risk of communication with freshwater aquifers, above which hydraulic fracturing is prohibited.

    A. described      B. allowed     C. prevented     D. forbidden

4. This requires both stringent regulations and a strong performance commitment by all companies involved in drilling and production-related activities to carry out operations to the highest possible standard.

    A. dedication     B. lie        C. promise      D. loyalty

5. Upgrading fluid-disposal systems so that storage and separation tanks replace open pits (closed-loop systems) can reduce the risk of accidental discharge of wastes during drilling.

    A. unexpected    B. deliberate   C. unintentional   D. unconscious

6. Conventional oil and gas developments generally follow a fairly well-defined sequence.

    A. accepted      B. traditional   C. well-known    D. popular

7. The blurred lines between the stages of an unconventional resource project development increase the complexity of the interactions between operator

and regulators.

    A. interfaces                 B. interpretations

    C. interventions              D. communications

    8. The regulatory system in most jurisdictions requires the <u>submission</u> and approval of a detailed field development plan at the end of the exploration phase.

    A. side          B. aspect         C. presentation      D. angle

    9. Relatively small <u>subsequent</u> alterations might trigger the need to resubmit and reapprove the entire development plan.

    A. preceding      B. following     C. substantial       D. substituted

    10. This broader <u>scope</u> is essential when it comes to assessments of water use and disposal and of future water requirements.

    A. extent         B. size           C. expanse          D. range

**III. Multiple choice：read the four suggested translations and choose the best answer.**

    1. <u>Sensitivity</u> at this stage to a range of above-ground concerns can do much to mitigate or avoid problems later in a development.

    A. 敏感         B. 积极         C. 焦虑          D. 乐观

    2. A careful study of the geology of the area <u>targeted</u> for drilling is necessary to allow operators to avoid operations in areas where deep faults or other characteristics create higher risks.

    A. 以……为榜样             B. 视……为朋友

    C. 视……为敌人             D. 以……为目标

    3. Such migration is <u>theoretically</u> possible in certain exceptional circumstances.

    A. 实际上地       B. 学术上地     C. 理论上地       D. 经验上地

    4. <u>Multiple</u> measures need to be in place to prevent leaks.

    A. 多种         B. 维护         C. 放逐          D. 远征

    5. For example，there are investments in <u>infrastructure</u> that may not proceed for an individual well but which would serve appreciably to reduce the cumulative environmental impacts of largescale development.

    A. 后勤         B. 基础设施     C. 项目           D. 结构

**IV. Put the following sentences into Chinese.**

    1. The choice of well site is a moment when engagement with local stakeholders and regulators needs to be handled with the utmost care.

2. A good understanding of the local geology and the use of micro-seismic (or other) measuring techniques for monitoring fractures is necessary to mini mise the residual risk.

3. Consider appropriate minimum-depth limitations on hydraulic fracturing to underpin public confidence that this operation takes place only well away from the water table.

4. At any given time an operator may be exploring or appraising part of a licence area, developing another part and producing from a third, with different regulatory approvals and permits applying at each stage.

5. Conventional oil and gas developments generally follow a fairly well-defined sequence, but the distinctions between the phases of an unconventional development can be much less clear-cut-development generally proceeds in a more incremental fashion.

**V. Put the following paragraphs into Chinese.**

1. Careful planning can greatly improve the productivity and recovery rates of wells, reducing the number of wells that need to be drilled and minimising the intensity of hydraulic fracturing and the associated environmental impact. Although the risk of triggering an earthquake is small, even minor earth tremors can easily undermine public confidence in the safety of drilling operations.

2. Governments are usually instrumental in promoting the co-ordinated and timely expansion of regional infrastructure alongside a gas development, ensuring that the regulatory framework serves to encourage or require the construction of gas transportation capacity or an expansion of local power supply. Strong co-ordination and communication is necessary between different branches and levels of government, as the rapid growth of a new industry puts pressure not only on the local physical infrastructure, but also on local social services.

# 7.2  Environmental Impact Reduction

## 中 Guidance to Reading

*An unavoidable impact of unconventional oil and gas extraction is a high land occupation. The major environmental disruptions are air emissions of pollutants, groundwater contamination due to uncontrolled gas or fluid blowouts or spills, leaking fracturing fluid, and waste water discharge. Fracturing fluids contain hazardous substances, and flow-back in addition contains heavy metals and radioactive materials from the deposit. Furthermore, groundwater contamination by methane, in extreme cases leading to explosion of residential buildings, and potassium chloride leading to salinization of drinking water is reported in the vicinity of gas wells. At a time when sustainability is key to future operations, the challenge for governments and regulators can be* **acute** *in relation to environmental impacts.*

## 🔲 Text

### Treat water responsibly

*Reduce freshwater use by improving operational efficiency; reuse or recycle, wherever practicable, to reduce the burden on local water resources.* Regulations covering shale gas, tight gas and coalbed methane production need to be designed to encourage operators to use water efficiently and to reuse and recycle it. The largest volumes of water are required for hydraulic fracturing: where the necessary economies of scale are present, it should be **feasible** to reuse and recycle significant volumes of the flow-back water from fracturing operations, reducing the issues and costs associated with truck traffic and with securing water supplies and wastewater disposal.

*Store and dispose of produced and waste water safely.* Rigorous and consistent regulations are needed to cover safe storage of waste water, with measures to ensure the robust construction and lining of open pits or, preferably, the use of storage tanks. Technology exists to treat produced and waste water to any standard, with the cost varying accordingly. It is the responsibility of regulators to set and enforce appropriate standards based on local factors, including the availability of freshwater supplies and options for disposal, without diminishing the operators' ultimate responsibility for operation in accordance with evolving best practice standards. The least-cost solution for producers may not be the most economically optimal solution, when the potential long-term benefits of using treated water and the wider social and environmental costs of discharges into water courses or evaporation ponds are taken into consideration.

*Minimise use of chemical additives and promote the development and use of more environmentally benign alternatives.* Disclosure of fracturing fluid additives can and should be **compatible** with continued incentives for innovation. The industry should commit to the development of fluid mixtures that, if they **inadvertently** migrate or spill, do not **impair** groundwater quality, or adopt techniques that reduce the need to use chemical additives.

### Eliminate venting, minimise flaring and other emissions

*Target zero venting and minimal flaring of natural gas during well completion and seek to reduce fugitive and vented greenhouse-gas emissions during the entire productive life of a well.* Best practice is to recover and market gas produced during the completion phase of a well, and public authorities need to consider **imposing** restrictions **on** venting and flaring and specific requirements for installing equipment to help minimise emissions. Measures in this area will also lower emissions of conventional pollutants, including **VOC**s. Operators should consider setting targets on emissions as part of their overall strategic policies to win public confidence that they are acting to minimise the environmental impact of their activities, taking into account the financial benefits of commercialising the gas that would otherwise be vented or flared. The gas industry as a whole, including conventional gas producers and companies operating in the midstream and downstream, needs to demonstrate that they are just as concerned by methane emissions beyond the production stage, for example in transportation and distribution.

*Minimise air pollution from vehicles, drilling rig engines, pump engines and compressors.* Pollution from vehicles and equipment is often controlled by existing environmental and **fuel-efficiency standards** (it is a responsibility of governments to ensure that appropriate standards are in place). Operators and service providers should consider the advantages of deploying the cleanest vehicles and equipment available, for example, electric vehicles and gas-powered rig engines, to reduce both local air and noise pollution.

### Be ready to think big

*Seek opportunities for realising the economies of scale and co-ordinated development of local infrastructure that can reduce environmental impacts.* Investments in infrastructure to reduce environmental impacts that may be commercially impossible to justify for an individual well can be justified for a larger development. Good regulation can help to realise these gains by ensuring appropriate **spatial** planning of licensing areas and of the associated infrastructure (such as access roads, water resources and disposal facilities, gas processing units, compression stations and pipelines). The concept of utility **corridors** and multi-use rights of way can be useful to concentrate infrastructure development and so limit the wider environmental impacts. Operators can realise these gains in various ways, for example by drilling **multiple wells** from a single **pad** (with horizontal bores tapping different parts of the reservoirs): this may result in greater **disruption** in the immediate vicinity of the site but can significantly reduce the wider environmental footprint. Another example is the construction of a pipeline network for water that requires **upfront investment** but **obviates** the need for many thousands of truck movements over the duration of a project and can lower unit costs. Good project and logistical planning by operators needs to go hand-in-hand with early strategic assessments and timely **interventions** by public authorities.

*Take into account the cumulative and regional effects of* **multiple drilling**, *production and delivery activities on the environment, notably on water use and disposal, land use, air quality, traffic and noise.* Development of any hydrocarbon resource involves a large amount of activity to build the necessary infrastructure, bring in supplies, drill wells, extract the resource, process it and transport it to market. This activity is enhanced for unconventional developments, because of the larger number of wells required. As a result, the level of activity that might be tolerable for individual wells, such as volumes of road traffic, land and water use or noise from drilling activity, can increase by orders of magnitude. Regulators need to assess the cumulative impact of these effects and respond appropriately. Assessment on a regional basis is particularly important in the case of water requirements.

The challenge for governments and regulators can be acute in relation to water resources and the risk of water contamination. Rigorous data collection, assessment and monitoring of water requirements (for shale and tight gas), and measurement of the quality of produced water (for coalbed methane) and of waste water (in all cases) are needed to allow **informed decisions** to be made.

## 🔲 Words and Expressions

| | |
|---|---|
| acute | 严重的;敏锐的;急性的;剧烈的 |
| feasible | 可行的;可能的 |
| compatible | 一致的;兼容的;适合的 |
| inadvertently | 不注意地;疏忽地;非故意地 |

| impair | 损害 |
| VOC | 挥发性有机化合物 |
| spatial | 空间的 |
| corridor | 走廊 |
| pad | 平台 |
| disruption | 分裂；破裂；毁坏 |
| obviate | 排除；消除；避免 |
| intervention | 介入；干预 |

## Phrases and Expressions

| impose on | 施加影响于 |
| fuel-efficiency standard | 能源效率标准 |
| multiple well | 多底井 |
| upfront investment | 先期投入 |
| multiple drilling | 丛式钻井 |
| informed decision | 合理的决定；知情的决定 |

## Language Focus

1. The largest volumes of water are required for hydraulic fracturing; where the necessary economies of scale are present, it should be feasible to reuse and recycle significant volumes of the flow-back water from fracturing operations, reducing the issues and costs associated with truck traffic **and** with securing water supplies and wastewater disposal.

（参考译文：用水量最大的是水力压裂过程：考虑到规模经济效益，应该切实可行地循环利用压裂作业中产生的大量返排水，减少卡车运水、水供应和废水处理等问题和成本。）

本句中，where 引导地点状语从句，it 是形式主语，动词不定式 to reuse and recycle significant volumes... 是真正的主语；reducing 引导的现在分词短语作 operation 的结果状语；reduce 的宾语是 the issues and costs，而 associated 引导的过去分词短语作后置定语修饰 the issues and costs，句中第三个 and 连接了两个 with 介词短语。

2. Operators should consider setting targets on emissions as part of their overall strategic policies to win public confidence that they are acting to minimise the

environmental impact of their activities, taking into account the financial benefits of commercialising the gas that would otherwise be vented or flared.

（参考译文：生产商应当把排放指标的制定视为其总体战略方针的一部分，使公众相信他们在努力减小其生产活动所带来的环境影响，并且将原本打算排放或燃烧掉的天然气卖出，获取一定的经济效益。）

本句中，setting targets on emissions 是 consider 的宾语，consider... as part of... 意思是"考虑把……作为……的一部分"，to win public... 是动词不定式表示目的，that 引导同位语从句，修饰 confidence，taking into account... 是现在分词短语做状语，这里的 that 从句作定语修饰 gas。

3. It is the responsibility of regulators to set and enforce appropriate standards based on local factors, including the availability of freshwater supplies and options for disposal, without diminishing the operators' ultimate responsibility for operation in accordance with evolving best practice standards.

［参考译文：监管者要根据当地具体情况，制定适当的标准（包括淡水供应及废水处理方法选择），并且监督执行。生产者要始终按照不断提升的标准对生产负责。］

本句中，it 作形式主语，真正的主语是动词不定式 to set and enforce appropriate standards；介词 including 作补语，进一步补充说明 standards，介词短语 without diminishing... 作伴随状语。

## ⊞ Reinforced Learning

### Ⅰ. Answer the following questions for a comprehension of the text.

1. What should be done to ensure the safe storage of waste water according to the text?

2. How can the use of chemical additives be minimized according to the text?

3. What is the best practice to realize zero venting and minimal flaring of natural gas during well completion according to the text?

4. How is pollution from vehicles and equipment often controlled?

5. What are the advantages and disadvantages of the construction of a pipeline network for water?

### Ⅱ. Multiple choice: choose the correct one from the alternative answers to give the exact meaning of the words.

1. Within an overarching performance framework, rigorous and consistent

regulations are needed to cover safe storage of waste water.

　　A. strict　　　　B. demanding　　　C. relentless　　　D. intense

　　2. Technology exists to treat produced and waste water to any standard, with the cost <u>varying</u> accordingly.

　　A. increasing　　B. fluctuating　　　C. wavering　　　D. changing

　　3. The least-cost solution for producers may not be the most economically <u>optimal</u> solution.

　　A. major　　　　B. best　　　　　　C. minor　　　　　D. popular

　　4. Operators and service providers should consider the advantages of <u>deploying</u> the cleanest vehicles and equipment available, for example, electric vehicles and gas-powered rig engines, to reduce both local air and noise pollution.

　　A. managing　　　B. installing　　　C. processing　　　D. working

　　5. Good regulation can help to realise these gains by ensuring <u>appropriate</u> spatial planning of licensing areas and of the associated infrastructure.

　　A. plenty of　　　B. enough　　　　C. proper　　　　D. a lotof

　　6. Development of any hydrocarbon resource involves a large amount of activity to build the necessary infrastructure, bring in supplies, drill wells, extract the resource, <u>process</u> it and transport it to market.

　　A. progress　　　B. make　　　　　C. treat　　　　　D. proceed

　　7. This activity is <u>enhanced</u> for unconventional developments, because of the larger number of wells required.

　　A. reinforced　　B. enriched　　　C. superior　　　D. diminished

　　8. Regulators need to assess the <u>cumulative</u> impact of these effects and respond appropriately.

　　A. accumulative　B. optional　　　C. additive　　　D. selective

　　9. Store and <u>dispose of</u> produced and waste water safely.

　　A. contain　　　B. deal with　　　C. transport　　　D. change

　　10. The gas industry as a whole, including conventional gas producers and companies operating in the midstream and downstream, needs to <u>demonstrate</u> that they are just as concerned by methane emissions beyond the production stage, for example in transportation and distribution.

　　A. display　　　B. protest　　　　C. show　　　　　D. examplify

**III. Multiple choice：read the four suggested translations and choose the best answer.**

1. Disclosure of fracturing fluid additives can and should be compatible with continued incentives for underlined innovation.

A. 创新　　　B. 改变　　　　C. 复制　　　　D. 移动

2. Investments in infrastructure to reduce environmental impacts that may be commercially impossible to justify for an individual well can be justified for a larger development.

A. 进步　　　B. 宣传　　　　C. 推进　　　　D. 影响

3. Good project and logistical planning by operators needs to go hand-in-hand with early strategic assessments and timely interventions by public authorities.

A. 前线的　　B. 后勤的　　　C. 人事的　　　D. 管理的

4. Regulators need to assess the cumulative impact of these effects and respond appropriately.

A. 研究　　　B. 评估　　　　C. 监督　　　　D. 保障

5. Existing users are deeply suspicious that their rights and water availability might be compromised.

A. 现有的　　B. 消失的　　　C. 将来的　　　D. 过去的

**IV. Put the following sentences into Chinese.**

1. The least-cost solution for producers may not be the most economically optimal solution，when the potential long-term benefits of using treated water and the wider social and environmental costs of discharges into water courses or evaporation ponds are taken into consideration.

2. The industry should commit to the development of fluid mixtures that，if they inadvertently migrate or spill，do not impair groundwater quality，or adopt techniques that reduce the need to use chemical additives.

3. The gas industry as a whole，including conventional gas producers and companies operating in the midstream and downstream，needs to demonstrate that they are just as concerned by methane emissions beyond the production stage，for example in transportation and distribution.

4. Pollution from vehicles and equipment is often controlled by existing environmental and fuel-efficiency standards（it is a responsibility of governments to ensure that appropriate standards are in place）.

5. Development of any hydrocarbon resource involves a large amount of activity to build the necessary infrastructure, bring in supplies, drill wells, extract the resource, process it and transport it to market.

**V. Put the following paragraphs into Chinese.**

1. Regulations covering shale gas, tight gas and coalbed methane production need to be designed to encourage operators to use water efficiently and to reuse and recycle it. The largest volumes of water are required for hydraulic fracturing: where the necessary economies of scale are present, it should be feasible to reuse and recycle significant volumes of the flow-back water from fracturing operations, reducing the issues and costs associated with truck traffic and with securing water supplies and wastewater disposal.

2. The challenge for governments and regulators can be acute in relation to water resources and the risk of water contamination. Rigorous data collection, assessment and monitoring of water requirements(for shale and tight gas), and measurement of the quality of produced water(for coalbed methane) and of waste water(in all cases) are needed to allow informed decisions to be made.

# 7.3　Social Responsibilities

## 🔲 Guidance to Reading

*Social responsibility is a duty every individual or organization has to perform so as to maintain a balance between the economy and the ecosystem. A trade-off always exists between economic development, in the material sense, and the welfare of the society and environment. Social responsibility means sustaining the equilibrium between the two. Corporate social responsibility functions as a built-in, self-regulating mechanism whereby a business ensures its active **compliance** with the spirit of the law, ethical standards, and international norms. The public authorities also need to impose restrictions on environmental unfriendly operations. The sustaining development of oil and gas industry requires action to be taken by governments, industry and private sector operators to earn and maintain public acceptance.*

## Text

### Ensure a consistently high level of environmental performance

*Integrate engagement with local communities, residents and other* **stakeholders** *into each phase of a development, starting prior to exploration; provide sufficient opportunity for comment on plans, operations and performance, listen to concerns and respond appropriately and promptly.*

Simply providing information to the public is not enough; both the industry and the public authorities need to engage with local communities and other stakeholders and seek the informed consent that is often critical for companies to proceed with a development. Operators need to explain openly and honestly their production practices, the environmental, safety, and health risks and how they are addressed. The public needs to gain a clear understanding of the challenges, risks and benefits associated with the development. The primary role of the public authorities in this context is to provide credible, science-based background information that can **underpin** an informed debate and provide the necessary **stimulus** for joint endeavour between the stakeholders.

*Establish baselines for key environmental indicators, such as groundwater quality, prior to* **commencing** *activity, and continue monitoring during operations.* This is a shared responsibility between the regulatory authorities, industry and other stakeholders. The data gathered needs to be made public and opportunities provided for all stakeholders to address any concerns raised, as an essential part of earning public trust. At a minimum, resource management or regulatory agencies must have groundwater quality information ( and, for coalbed methane production, information on groundwater levels) in advance of new drilling activities, so as to provide a baseline against which changes in water level and quality can be compared.

*Measure and disclose operational data on water use, on the volumes and characteristics of waste water and on methane and other air emissions, alongside full,* **mandatory** *disclosure of fracturing fluid additives and volumes.* Good data, measurement and transparency are vital to public confidence. For example, effective tracking and documentation of waste water is necessary to **incentivise** and ensure its proper treatment and disposal. **Reluctance** to disclose the chemicals used in the hydraulic fracturing process and the volumes involved, though understandable in terms of commercial competition, can quickly breed mistrust among local citizens and environmental groups.

*Minimise disruption during operations, taking a broad view of social and environmental responsibilities, and ensure that economic benefits are also felt by local communities.* Existing legislation and regulations usually require operators to act in an environmentally and socially responsible manner, but operators need to go beyond minimally satisfying legal requirements in demonstrating their commitment to local development and environmental protection, for example through attention to local concerns about the volume and timing of truck traffic. Particularly in jurisdictions where mineral rights are owned by the state (rather than as in parts of the United States, where surface landowners might also be **subsurface** mineral rights holders, entitled to royalty payments), it is essential that tangible benefits are evident at the local level, where production occurs. This can be difficult to achieve in a timely manner, given the delay between the start of a development project and the moment at which revenues start to flow, whether to government, the mineral rights' owner or the operator. Early public commitment by authorities and developers to expand local infrastructure and services **in step with** exploration and production activities can help. Governments need to be willing to consider using part of the revenues (from taxes, royalties, etc.) to invest in the development of the areas in question.

*Ensure that anticipated levels of unconventional gas output are matched by commensurate resources and political backing for robust regulatory regimes at the appropriate level, sufficient permitting and compliance staff, and reliable public information.* An important focus for governments should be on ensuring there is a sufficient knowledge base on all environmental and technical aspects of unconventional gas development, that high-quality data are available and that sound science is being applied and promoted. Well-funded, suitably skilled and motivated regulators, in sufficient numbers, are essential to the responsible development of an unconventional resource.

*Find an appropriate balance in policy-making between prescriptive regulation and performance-based regulation in order to guarantee high operational standards while also promoting innovation and technological improvement.* In some areas, detailed rules and checks are **indispensable** to guarantee environmental performance; but it is not always possible, or desirable, to regulate every aspect of a process in which technology is moving rapidly. Setting performance criteria and allowing operators to find the best way to meet them can often provide a better outcome than a **prescriptive** approach. Examples of performance criteria might be a mandated minimum level of improvement in water usage or a requirement that a "best-in-class" cement quality measurement is run, the burden being on the operator to prove the use of best-inclass.

Whichever approach or combination of methods is chosen, there needs to be strict **enforcement** and **penalties** in the case of non-compliance, ultimately including loss of the **licence to operate**.

*Ensure that emergency response plans are robust and match the scale of risk.* Operators and local emergency services should have robust plans and procedures in place to respond quickly and effectively to any accident, including appropriate training and equipment.

*Pursue continuous improvement of regulations and operating practices.* Technology and best practice are constantly evolving. While respecting the advantages of clarity and stability in regulation, governments must be ready to **incorporate** lessons learned from experience in a dynamic industrial sector. For industry, following best practice means constant readiness to raise standards and providing the means to meet them.

*Recognise the case for independent evaluation and verification of environmental performance.* Credible, third-party certification of industry performance can provide a powerful tool to earn and maintain public acceptance, as well as providing a powerful tool to assist companies to **adhere** to best practices. These independent assessments should come from institutions that enjoy **public trust**, whether academic or research institutes or independent regulatory or certification bodies.

Application of these Golden Rules requires action to be taken by both governments and industry. While the ultimate responsibility for sustaining public confidence rests with the industry, it is governments that need to set the regulatory framework, **promulgate** the required principles and provide support through many related activities, *e. g.* scientific research. Trying to **specify** precisely the roles of governments, gas producers and other private sector operators in each area is not practicable on a global scale. Conditions vary from country to country, including the legal, geological, social and political background, land-use practices, water availability and many others.

## 🔲 Words and Expressions

| | |
|---|---|
| compliance | 认证 |
| stakeholder | 股东 |
| underpin | 从下头支持;支撑 |
| stimulus | 刺激;刺激品;激励 |
| commence | 开始 |
| mandatory | 命令的,强制性的 |
| incentivize | 刺激;激励 |
| reluctance | 不愿;勉强 |
| subsurface | 地下的;表面下的 |
| indispensable | 不可缺少的 |
| prescriptive | 规定的;(语言学)规范的;约定俗成的 |
| enforcement | 执行;强制;实施 |
| penalty | 处罚 |
| incorporate | 包含;吸收 |
| adhere | 遵守;坚持;粘附 |
| promulgate | 发布;颁布;发表 |
| specify | 详细说明;指定;阐述 |

## 🔲 Phrases and Expressions

| | |
|---|---|
| in step with | 与……一致 |
| licence to operate | 经营许可证 |
| public frust | 公信度,公众信任 |

## 🔲 Language Focus

1. The primary role of the public authorities in this context is to provide credible, science-based background information that can underpin an informed debate and provide the necessary stimulus for joint endeavour between the stakeholders.

（参考译文:公共管理部门的主要职能则是提供可靠的、科学性的背景信息,利用这些信息可以进行合理辩论,并激励股东之间的携手合作。）

本句中,that 引导定语从句,修饰 background information。定语从句中有两个并列的谓语动词：underpin 和 provide。

2. Particularly in jurisdictions where mineral rights are owned by the state (rather than as in parts of the United States, where surface landowners might also be **subsurface** mineral rights holders, entitled to royalty payments), it is essential that tangible benefits are evident at the local level, where production occurs.

［参考译文：特别是在矿权归州所有的管辖区内（而不像美国的某些地区，地面土地所有者也可能是地下矿权持有者，有权获得特许使用费），必须要让生产区居民获得实实在在的好处。］

在本句中，where mineral rights are owned by the state 作定语修饰 jurisdictions。rather than 为并列连词，连接 owned by the state 和 entitled to royalty payments 两个并列成分。as 作连词，意思是"像……一样"。

3. Examples of performance criteria might be a mandated minimum level of improvement in water usage or a requirement that a "best-in-class" cement quality measurement is run, the burden being on the operator to prove the use of best-in-class.

（参考译文：例如对水利用改进方面设定了最低标准，或者是要求检测最佳水泥质量。这对施工方是一种压力，他们必须证明使用了最佳水泥。）

本句中，or 连接两个并列的表语 a mandated minimum level of improvement 和 a requirement。that 引导同位语从句修饰 requirement；the burden being... 为独立主格结构作状语，进行补充说明。

4. These independent assessments should come from institutions that enjoy public trust, whether academic or research institutes or independent regulatory or certification bodies.

（参考译文：这些独立评估应该来自于受公众信任的机构，无论是学术或研究机构或是独立的监管及认证机构。）

本句中，that enjoy public trust 是定语从句，修饰 institutions；whether... or... 是关联词语，意思是"无论（不管）……还是……"。

## ⬡ Reinforced Learning

### I. Answer the following questions for a comprehension of the text.

1. What do operators have to do to ensure a consistently high level of environmental performance?

2. What isvital to public confidence when talking about water use and disposal of waste water?

3. What is the primary role of the public authorities to ensure a consistently high level of environmental performance?

4. What do existing legislation and regulations usually require operators to do?

5. Why is it not always possible to regulate every aspect of a process?

**II. Multiple choice: choose the correct one from the alternative answers to give the exact meaning of the words.**

1. Operators need to explain openly and honestly their production practices, the environmental, safety, and health risks and how they are addressed.

A. truthfully      B. shyly      C. lovely      D. really

2. This is a shared responsibility between the regulatory authorities, industry and other stakeholders.

A. obedience      B. obligation      C. choice      D. option

3. Good data, measurement and transparency are vital to public confidence.

A. good      B. essential      C. healthy      D. helpful

4. Reluctance to disclose the chemicals used in the hydraulic fracturing process and the volumes involved, though understandable in terms of commercial competition, can quickly breed mistrust among local citizens and environmental groups.

A. take      B. increase      C. defeat      D. foster

5. An important focus for governments should be on ensuring there is a sufficient knowledge base on all environmental and technical aspects of unconventional gas development, that high-quality data are available and that sound science is being applied and promoted.

A. almost      B. hungry      C. enough      D. huge

6. For example, effective tracking and documentation of waste water is necessary to incentivise and ensure its proper treatment and disposal.

A. efficient      B. valid      C. exact      D. struggling

7. In some areas, detailed rules and checks are indispensable to guarantee environmental performance.

A. extended      B. specific      C. exchanged      D. shortened

8. While respecting the advantages of clarity and stability in regulation, governments must be ready to incorporate lessons learned from experience in a dynamic industrial sector.

A. purification             B. constructiveness

C. confidence              D. clearness

9. Credible, third-party certification of industry performance can provide a powerful tool to earn and maintain public acceptance, as well as providing a powerful tool to assist companies to <u>adhere</u> to best practices.

A. comply      B. conform      C. transform      D. refer

10. <u>Application</u> of these Golden Rules requires action to be taken by both governments and industry.

A. Optimism      B. Using      C. Performance      D. Tragedy

**Ⅲ. Multiple choice: read the four suggested translations and choose the best answer.**

1. The primary <u>role</u> of the public authorities in this context is to provide credible, science-based background information that can underpin an informed debate and provide the necessary stimulus for joint endeavour between the stakeholders.

A. 猜测      B. 观念      C. 角色      D. 假想

2. This is a shared <u>responsibility</u> between the regulatory authorities, industry and other stakeholders.

A. 利润      B. 责任      C. 关系      D. 基金

3. This can be difficult to <u>achieve</u> in a timely manner, given the delay between the start of a development project and the moment at which revenues start to flow, whether to government, the mineral rights' owner or the operator.

A. 达到      B. 报答      C. 开除      D. 想象

4. Governments need to be willing to consider using part of the revenues (from taxes, royalties, etc.) to <u>invest</u> in the development of the areas in question.

A. 装修      B. 维护      C. 建立      D. 投资

5. Whichever approach or combination of methods is chosen, there needs to be strict enforcement and <u>penalties</u> in the case of non-compliance, ultimately including loss of the licence to operate.

A. 奖赏      B. 批评      C. 处罚      D. 鼓励

**Ⅳ. Put the following sentences into Chinese.**

1. Simply providing information to the public is not enough; both the industry and the public authorities need to engage with local communities and other

stakeholders and seek the informed consent that is often critical for companies to proceed with a development.

2. The primary role of the public authorities in this context is to provide credible, science-based background information that can underpin an informed debate and provide the necessary stimulus for joint endeavour between the stakeholders.

3. The data gathered needs to be made public and opportunities provided for all stakeholders to address any concerns raised, as an essential part of earning public trust.

4. This can be difficult to achieve in a timely manner, given the delay between the start of a development project and the moment at which revenues start to flow, whether to government, the mineral rights' owner or the operator.

5. An important focus for governments should be on ensuring there is a sufficient knowledge base on all environmental and technical aspects of unconventional gas development, that high-quality data are available and that sound science is being applied and promoted.

**V. Put the following paragraphs into Chinese.**

1. Good data, measurement and transparency are vital to public confidence. For example, effective tracking and documentation of waste water is necessary to incentivise and ensure its proper treatment and disposal. Reluctance to disclose the chemicals used in the hydraulic fracturing process and the volumes involved, though understandable in terms of commercial competition, can quickly breed mistrust among local citizens and environmental groups.

2. Application of these Golden Rules requires action to be taken by both governments and industry. While the ultimate responsibility for sustaining public confidence rests with the industry, it is governments that need to set the regulatory framework, promulgate the required principles and provide support through many related activities, *e. g.* scientific research. Trying to specify precisely the roles of governments, gas producers and other private sector operators in each area is not practicable on a global scale. Conditions vary from country to country, including the legal, geological, social and political background, land-use practices, water availability and many others.

# Chinese Translation and Key to Exercises

# 第1章　非常规天然气介绍

## 1.1　非常规天然气资源

### 导语

　　一个被广为接受的非常规油气的定义是非常规油气是以非常规手段开发的油气。从化学成分上来看，非常规油气和常规油气是相同的。常规石油包括原油、天然气凝析液和凝析油；而非常规石油种类更多，包括油砂、超重油和页岩油，但需要先进技术开采。非常规天然气存于高度致密的岩石或煤层中，需要使用特殊开采技术才能得到。总之，非常规油气开采不易且成本高。随着经济和技术的发展，目前认为是非常规的油气资源也会归入常规类别中的。"常规"和"非常规油气"种类不会是一成不变的。

### 课文

　　非常规天然气是天然气资源中的一部分，人们一直认为其开发难度大、成本高。本文将重点涉及三种主要的非常规天然气：页岩气、致密气和煤层气。

　　页岩气是一种常见岩石即页岩中的天然气。页岩层的特点是渗透率低，与常规油藏相比，气体在页岩中的流动能力较差。页岩层富含有机质，与大部分油藏不同的是，页岩层是典型的天然气生油母质，也就是说页岩气是吸附在生油岩表面或聚集在其内部的天然气。

　　煤层气，也称为煤层甲烷，是煤层中的天然气。最初开采煤层甲烷是为了使矿场更安全，但现在大部分是从不可开采煤层中生产了。

　　致密气是对存在于低渗透率地层中的天然气的通称。这些低渗透率气藏如果不使用水力压裂等增产措施使气体加速流向井底，就不能使生产经济可行。致密气没有明确分类，与常规气、页岩气之间的界限也不甚清晰。

　　尽管非常规天然气的开发周期和生产技术与上游工业中其他类生产有很多共同点，但非常规天然气开发确实有其不同的特性和要求，尤其明显的是所面临的环境破坏风险较高，以及由此带来的不良社会影响。这就解释了为什么非常规天然气开采引发了如此多的争议。

非常规天然气开采会给环境带来较大的潜在影响,主要原因在于资源本身的特性:非常规资源比常规资源的沉积更加分散,而且不易开采。非常规天然气聚集在致密或低渗透率的岩石中,流动阻碍较大,因此难以开采。由于这些资源比较分散、难以开采,所以要生产一定量的非常规天然气,其生产规模要远大于常规天然气开发。这就意味着其钻井和生产活动更加具有侵害性,并将造成更大的环境问题。

非常规天然气开采需要更大范围的生产空间,其中的一个明显特征就是所需油气井数更多。陆上常规油田每十平方千米最多需要一口井,而非常规油田每一平方千米可能就至少需要一口井,这就使钻井和完井工程对环境和当地居民的影响大大增加。

另外重要的一点是,非常规天然气开采还需要做更复杂且强度更大的准备工作。水力压裂技术虽然有时也用于常规油藏的增产,但致密气和页岩气开发几乎总是需要使用这种技术,使流体以足够快的速度流向井底。水力压裂还可用于开采煤层气,但较之上述气藏,其使用频率较低。该技术应用过程中,水的使用和排放会引起很多环境问题,如淡水资源枯竭,地表水和地下水可能污染等问题。

非常规天然气的开采还会导致大气中温室气体浓度增加,并影响当地空气质量。在某些情况下也会导致空气中工业排放物增加,如甲烷(一种强温室气体)、挥发性有机物(可致使烟雾形成)和二氧化碳(开采过程中大量能耗的产物,比常规开采能量消耗更大)。但是这些危害到底比常规开采大多少是不确定的,很大程度上取决于开采方法的选择。另一方面,非常规天然气的开采会带来潜在的净收益,从这个意义上讲,严格按照环境标准进行开采和运输的天然气将被广为使用,取代含碳量更高的煤和石油。

非常规油气藏在单位储层空间中的可采储量较小,因此开发所占地的面积就要大得多。此外,非常规油气储量丰富的地区现代工业实力并不一定都很强;该地区的常规油气储量也不一定很丰富,某些地区可能几乎没有油气产量或最近没有。这种情况会降低公众的认可程度。

**非常规天然气开采和地震风险**

由开采非常规天然气而引发的地震已经时有发生,例如 Cuadrilla 公司在英国黑潭附近的页岩气开采案例,美国俄亥俄州的扬斯敦案例,后者已初步确定与废水的注入有关,废水注入在某些方面与水力压裂很相似。记录中的是只有 2 级的小型地震,可以为人类所感知,但不会造成任何地表损害。

水力压裂会使地层深部的岩石产生裂缝,导致经常发生一些小型地震。

实际上石油工程师通过这些小地震可以对水力压裂的过程进行监测。一般来说这些小型地震由于震级过小，在地面无法检测到，所以在监测过程中需要用到特殊的观察井和十分灵敏的仪表。当井筒或裂缝发生交切时会激活已有断层，产生较大地震。Cuadrilla 案例中的情形似乎就是如此。

　　水力压裂并不是唯一可以引发小型地震的人为手段。任何促使地下应力产生的活动都有引发地震的风险。据报导某些大型建筑和大坝的施工已经引发了小型地震。

　　因为冷水循环到地下进入地热井中，所以地热井产生热诱导应力，足以引发人类可以感觉到的地震。这种情况同样发生于深井开采的过程中。在进行非常规天然气开发时，一定要对当地的地质情况进行仔细勘探，判断是否存在深部断层及其他危险性的地质特征，避免在具有危险地质特征的区域进行压裂。在任何情况下，监测都是必不可少的，有了监测机制，发现地震活动逐渐加强时，就可以停止相关操作。

## Key to Exercises

Ⅰ. 1. Unconventional gas refers to a part of the gas resource base that has traditionally been considered difficult or costly to produce.

2. The nature of the recourses is the major reason. Firstly, unconventional resources are less concentrated than conventional deposits and do not give themselves up easily. What's more, since the resources are more diffuse and difficult to produce, they need much more operations which can be considerably more invasive, involving a generally larger environmental footprint.

3. (1) It needs more wells to extract unconventional gas.

(2) The technique of hydraulic fracturing gives rise to a number of environmental concerns, including depletion of freshwater resources and possible contamination of surface water and aquifers.

(3) The production of unconventional gas also contributes to the atmospheric concentration of greenhouse gases and affects local air quality.

4. More wells are required to extract unconventional gas. While onshore conventional fields might require less than one well per ten square kilometers, unconventional fields might need more than one well per square kilometer, significantly intensifying the impact of drilling and completion activities on the environment and local residents.

5. First, careful survey should be done on the geology of the area to be

developed, and avoid hydraulic fracturing in the areas with deep faults or other enhanced risk. Second, monitoring is necessary so that operations can be suspended if there are signs of increased seismic activity.

Ⅱ. 1. A  2. D  3. B  4. A  5. B  6. A  7. C  8. B  9. B  10. A

Ⅲ. 1. C  2. A  3. B  4. A  5. A

Ⅳ. 1. 这些低渗透率气藏如果不使用水力压裂等增产措施使气体加速流向井底,就不能使生产经济可行。

2. 非常规天然气聚集在致密或低渗透率的岩石中,流动受到阻碍,因此难以开采。

3. 非常规天然气的开采需要更大规模的生产操作,其特点之一就是所需井数更多。

4. 记录中的是只有 2 级左右小型地震,虽可以为人类所感知,但不会造成任何地表损害。

5. 在任何情况下,监测都是必不可少的。有了监测机制,地震活动有加强的迹象时,就可以停止相关操作。

Ⅴ. 1. 非常规油气藏在单位储层空间中的可采储量较小,因此开发所占地理面积就要大得多。此外,非常规油气储量丰富的地区现代工业实力并不一定都很强;该地区的常规油气储量也不一定很丰富,某些地区可能几乎没有油气产量或最近没有。这种情况会降低公众的认可程度。

2. 水力压裂会使地层深部的岩石产生裂缝,导致经常发生一些小型地震。实际上石油工程师通过这些小地震可以对水力压裂的过程进行监测。一般来说这些小型地震由于震级过小,在地面无法检测到,所以在监测过程中需要用到特殊的观察井和十分灵敏的仪表。当井筒或裂缝发生交切时会激活已有断层,产生较大地震。

# 1.2　页岩气

## 导语

页岩气蕴藏量极大,是世界可采非常规天然气的主力。但页岩渗透率低,不足以让大量流体流入井筒,所以大部分页岩不适于大量生产天然气。然而,美国页岩气发展迅猛,越来越多的天然气从页岩中开采出来。由于应用了水力压裂和水平钻井技术,页岩气新资源的开发弥补了常规气藏产量的下降,使美国天然气储量有了大幅上升。

## 课文

全球页岩气资源储量丰富,据估计超过 $18,775 \times 10^{12} \, \text{ft}^3$,页岩气资源储量或许超过了世界所有常规资源的总量。目前全球天然气年消耗量约为 $110 \times 10^{12} \, \text{ft}^3$,也就是说按照当前消耗水平,即使能开采出页岩气资源总量的 25%,就足以满足全球 40 多年的天然气消耗需求。由此可见全球页岩气资源总量之丰富。

含气页岩是富含有机质的地层,总有机碳重量占其成分高达 25%,一般埋藏于地下 3~6km 处,远深于常规气藏。页岩基质渗透率较低,因此,页岩既是烃源岩也是储集岩。另外,页岩在地下成水平走向,大大增加了页岩气勘探和开采的难度与成本,因此不同于常规天然气资源。

**页岩气资源的地质特征**

页岩是一种沉积岩,主要由黏土大小的颗粒固化形成。在低流动的沉积环境下,如潮滩和深水盆地等,细粒黏土会沉积成泥,形成页岩。同时,藻类、植物和动物等有机残骸往往也会在沉积过程中圈闭于其中。

随着沉积物的积聚,原本板状黏土颗粒往往会形成扁平结构。沉积多年之后,这些沉积物高度压实与固化后形成层状页岩。压实作用和层状结构导致页岩层渗透率极低。数百万年之后,聚集在页岩中的有机物质在高温高压作用下形成碳氢化合物,如天然气等。图 1.1 显示了典型的页岩露头,可以看到天然相嵌的水平页岩层和近乎垂直切断水平层的天然裂缝。

图 1.1　页岩露头

天然气有的圈闭在矿物颗粒之间的孔隙里,有的在天然裂缝中,还有的因化学原理吸附在页岩有机质的表面。页岩基质渗透率较低,因此岩石内聚集的气体无法轻易流动,通常页岩渗透率在 $0.00001 \sim 0.01 \text{mD}$。

因此,页岩气既生成于页岩层内又储存于页岩层内,这种地层既是天然气的气源岩又是储集岩,常称为含气页岩。

**页岩气的化学组成**

天然气是碳氢化合物,主要成分是甲烷,另有少量的乙烷、丙烷、丁烷和其他气体。天然气无色无味,燃烧时释放巨大能量。把大部分通常伴生的碳氢化合物去除后,天然气几乎就是纯净的甲烷了,这种加工后的天然气称为"干气"。含有其他碳氢化合物的天然气称为"湿气"。尽管一些页岩地层也会产生湿气,但一般的页岩气为含甲烷在 90% 及以上的干气。

采出气经过加工,去除一种或多种成分后,可供民用和工业应用。通常去除的气体组分包括硫化氢、二氧化碳、氮气、重烃和水蒸气等,其目的在于达到管道输送、安全、环保和质量等方面的相关标准。加工方法根据所需去除的组分和气流的相关性能(温度、压力、组成、流速等)而定。因为页岩气组成不同,各矿区对页岩气的加工要求也不同,甚至在同一矿区对不同井产出的页岩气加工要求也不一样。

**经济可采的天然气资源所具有的共同特征**

具有经济可采储量的页岩有一些共同特征。这些储层一般是:

● 有机质含量丰富,总有机碳量可高达其总重量的 25%。

● 在地下 3~6km 的成熟烃源岩,温度在 100~200℃(又称为热成因气窗口);在这一深度下,高温和高压使有机质转化成了天然气。

● 具有足够脆性和刚性,能产生裂缝并保持裂缝张开。

在经济可采盆地内及这些盆地之间,含气页岩都有其各自的特殊性。例如,热成熟度和化学组分各有不同,因此,有的页岩所含甲烷量较理想,有的烃气体和液体相对分子质量较高。页岩地层内和地层之间的含水量、油藏压力、厚度、渗透率及孔隙度等也存在差异。是否存在如断层这样的地质特征是区分页岩的另一个指标。当评估某个气藏的产能和研究钻进方案时,地质学家和工程师必须将所有相关因素考虑在内,才能做出正确的选择。

在过去的 20 年里,勘探、钻井、开采技术等方面均已取得巨大进步,使得低渗透率页岩气评估和开采更加经济可行。水平钻井技术的进步提高了钻井效率以及页岩气开采效率。这些技术可以使气井到达产层并沿狭长的水平产层精确地延伸。这不仅有利于接触油藏,而且还能进入之前难以到达的油气藏,例如处于人口稠密的居住区下面或者湖泊下面的气藏。

油藏增产措施(特别是水力压裂)的发展,进一步促进了低渗透页岩中气体的流动。这个过程本质上是通过控制水力压裂来产生导流通道让气体流入井筒。经过逐步改进,压裂液已经可以应对各种类型页岩(从低压、水敏性黏土页岩到高压脆性页岩)的挑战了。多级压裂工具的出现也使长距离水平段压裂效率有所提高。

🖵 **Key to Exercises**

Ⅰ.1. The current global natural gas consumption is estimated at about 110 Tcf annually.

2. Shales have low permeability matrices and act as both the source and reservoir rock for gas.

3. The combination of compaction and lamination result in a formation with extremely limited permeability.

4. Natural gas is a combination of hydrocarbon gases consisting primarily of methane($CH_4$)and,to a lesser extent,butane,ethane,propane,and other gases.

5. These technologies enable wells to reach and accurately follow long,narrow,horizontal pay streaks. This not only promotes increased reservoir contact, but also provides access to gas reservoirs that were previously difficult or impossible to reach.

Ⅱ.1. A  2. C  3. B  4. A  5. A  6. B  7. D  8. A

Ⅲ.1. B  2. A  3. D  4. A  5. B

Ⅳ.1. 按照当前消耗水平,即使能开采出页岩气资源总量的25%,就足以满足全球40多年的天然气消耗需求。

2. 数百万年之后,聚集在页岩中的有机物质在高温高压作用下形成碳氢化合物,如天然气等。

3. 多年沉积之后,这些沉积物高度压实与固化后形成层状页岩。

4. 是否存在如断层这样的地质特征是区分页岩的另一个指标。

5. 这个过程实质上是通过控制水力压裂来产生导流通道让气体流入井筒。

Ⅴ.1. 气页岩是富含有机质的地层,总有机碳重量占其成分高达25%,一般生存于地下3～6km处,远深于常规气藏。页岩基质渗透率较低,因此,页岩既是烃源岩也是储集岩。另外,页岩在地下成水平走向,大大增加了页岩气勘探和开采的难度与成本,因此不同于常规天然气资源。

2. 在过去的20年里,勘探、钻井、开采技术等方面均已取得巨大进步,使得低渗透率页岩气评估和开采更加经济可行。水平钻井技术的进步提高了钻井效率以及页岩气开采效率。这些技术可以使气井到达产层并沿狭长的水平产层精确地延伸。这不仅有利于接触油藏,而且还能进入之前难以到达的油气藏,例如处于人口稠密的居住区下面或者湖泊下面的气藏。

# 1.3  焦油砂

## 🔲 导语

油砂、焦油砂或更专业点地说沥青砂是一种非常规石油沉积物。世界上很多石油是以油砂的形式存在,只不过是不能全部开采出来罢了。只是到最近油砂才被认可是世界石油储量的一部分,因为高油价和新技术使油砂的开采和加工具经济可行性。然而,焦油砂最不利于环境,因为开采和加工都需要大量能源,焦油砂生产排放的温室气体是常规石油开采的三倍。

## ⌐ 课文

焦油砂(也称为油砂)是一种黏土、砂、水和沥青(一种重质黑色稠油)的混合物,如图1.2所示。通过开采和加工焦油砂,可获取富含石油的沥青质,然后再将沥青质炼制成石油。油砂中的天然沥青无法从地下泵上来,但油砂沉积物可以开采出来,通常使用露天开采技术或地下加热处理技术开采。

焦油砂生产出的石油与常规油井开采出来的石油相似,但前者比后者的过程更加复杂,如图1.3所示。油砂开采加工包括开采系统和分离系统,将焦油砂中的沥青质和黏土、砂及水分离。为了提取油砂中的沥青质,首先要将矿石与温水混合制成泥浆;然后将泥浆注入到处理器中,沥青质便可从水和砂的混合物中分离出来;沥青质需要再加工后才能炼制。由于沥青质具有高黏性,所以需要用轻烃进行稀释,以便可以用管道运输。提取出的沥青质要用一种特殊溶剂进行稀释,然后通过管道运输至精加工工厂,转变为许多不同类别的优质低硫、低黏度合成原油。

图1.2　焦油砂

图1.3　焦油砂生产

目前,美国的焦油砂尚未到达较高的商业化开采水平。实际上,只有加拿大焦油砂产业规模很大,在进行商业化生产。不过委内瑞拉也有少量焦油砂在进行商业化生产。

加拿大的焦油砂产业集中在艾伯塔,每天可以产出一百多万桶的合成油。目前,加拿大油砂油产量占其石油总产量的40%左右,且正在迅速增长。美国大约20%的原油和成品油来自加拿大,而且绝大部分都出自焦油砂。焦油砂可以通过采矿提取,也可以原地提取。加拿大的焦油砂与美国的有所不同,前者亲水,后者亲烃。由于这种差异性,犹他州焦油砂的提取技术不同于艾伯塔的技术。

最近,原油价格再次上涨,可能会促使美国进行焦油砂商业化生产,政府和石油企业部门都对焦油砂资源开发表示了兴趣,欲将其替代常规石油资源。

**焦油砂开采和加工**

近地表的焦油砂沉积物,可以通过露天采矿技术进行开采。20 世纪 90 年代引进的新方法大大提高了焦油砂的开采效率,从而降低了成本。这些方法包括使用大型液压挖土机和电动挖土机挖掘焦油砂,然后装载到巨型卡车上,这种卡车每次可负载多达 320t 的焦油砂,如图 1.4 所示。

焦油砂挖出后,运至提炼厂,经过热水处理,沥青质便与砂、水以及矿物质分离开来,该分离过程一般在分离间进行,如图 1.5 所示。焦油砂加高温热水后产生泥浆,泥浆通过管道运至提炼厂,进行搅拌。经过热水混合搅拌后,沥青就从焦油砂中释放出来;微小气泡附着在沥青质微滴的表面,使沥青质微滴漂浮到分离器的上部,从而将沥青质撇出去;然后再去掉残余水和固体残留物,将沥青质运走,最终精炼成合成原油。

图 1.4　加拿大艾伯塔的焦油砂露天矿场　图 1.5　加拿大艾伯塔的焦油砂提取分离车间

生产一桶原油大约需要两吨焦油砂,焦油砂大约可以提取 75% 的沥青质。采油结束后,废弃油砂及其他物质都要运回矿场,使矿场恢复再生。

现场开采法适用于埋藏较深不宜开采,且开采不经济的沥青质沉积物。现场生产技术包括注蒸汽法、注溶剂法和火烧油层法(注入氧气,燃烧部分油砂提供热量)。目前为止注蒸汽法最受青睐。在这些提取方法中,有的需要大量的水和能量(用于加热和泵送)。

焦油砂开采与加工都会对环境产生各种影响,如全球变暖、温室气体排放、土地破坏、野生动物生存影响、空气和水质量降低等。美国商业化焦油砂生产的发展给当地社区业带来了巨大的社会和经济影响。美国西部地区相对干旱,油砂加工需要耗费大量的水,因此引起了人们的特别关注;目前,每生产一桶原油,就需要好几桶水来提取和加工焦油砂,不过其中有一部分水可以回收。

焦油砂这项"庞大工程"是人类历史上规模最大的工程,也可能是最具破坏性的工程。焦油砂开采过程中,二氧化碳的排放量至少是常规石油开采的三倍,这也是造成北美气候变化的主要因素。

焦油砂的开采导致地球上森林覆盖率迅速降低,其速度仅次于亚马逊雨林盆地的森林砍伐速度。从当前已批准的项目看,到2018年每天将生产 $300 \times 10^4$ bbl 焦油砂原油;每桶油的生产要耗费多达5bbl的水。

许多地区的人类健康状况已经在逐步恶化,据称很多都由焦油砂的开采所引起。焦油砂的开采给艾伯塔地区带来了许多严重的社会问题,从住房危机到临时外国劳工项目(歧视并剥削所谓的非本国公民)的疯狂扩张。石油运输管道、海上超级油轮以及炼油厂等基础设施从四面八方穿越北美大陆通向三大洋和墨西哥湾。

## Key to Exercises

I. 1. Tar sands (also referred to as oil sands) are a combination of clay, sand, water, and bitumen, a heavy black viscous oil.

2. The bitumen in tar sands cannot be pumped from the ground in its natural state; instead tar sand deposits are mined, usually using strip mining or open pit techniques, or the oil is extracted by underground heating with additional upgrading.

3. These techniques include steam injection, solvent injection, and firefloods, in which oxygen is injected and part of the resource burned to provide heat.

4. Both mining and processing of tar sands involve a variety of environmental impacts, such as global warming and greenhouse gas emissions, disturbance of mined land; impacts on wildlife and air and water quality.

5. Canadian tar sands are different than U. S. tar sands in that Canadian tar sands are water wetted, while U. S tar sands are hydrocarbon wetted.

II. 1. A   2. C   3. D   4. A   5. B

III. 1. B   2. A   3. C   4. B   5. D

IV. 1. 通过开采和加工焦油砂,可获取富含石油的沥青质,然后再将沥青质炼制成石油。

2. 焦油砂生产出的石油与常规油井开采出来的石油相似,但前者比后者的过程更加复杂。

3. 采油结束后,废弃油砂及其他物质都要运回矿场,使矿场恢复再生。

4. 现场生产技术包括注蒸汽法、注溶剂法和火烧油层法(注入氧气,燃烧部分油砂提供热量)。

5. 许多地区的人类健康状况已经在逐步恶化,据称很多都由焦油砂的开采所引起。

Ⅴ.1. 近地表的焦油砂沉积物,可以通过露天采矿技术进行开采。20 世纪 90 年代引进的新方法大大提高了焦油砂的开采效率,从而降低了成本。这些方法包括使用大型液压挖土机和电动挖土机挖掘焦油砂,然后装载到巨型卡车上,这种卡车每次可负载多达 320 吨的焦油砂。

2. 焦油砂的开采导致地球上森林覆盖率迅速降低,其速度仅次于亚马逊雨林盆地的森林砍伐速度。从当前已批准的项目看,到 2018 年每天将生产 300 万桶焦油砂原油;每桶油的生产要耗费多达 5 桶的水。

# 1.4 天然气凝析液

## 🔲 导语

天然气凝析液也是碳氢化合物,是从天然气中分离出来的液体。分离出的天然气凝析液就贵重了,因此从天然气中脱除凝析液有利可图。脱除的凝析液还可以再分离成不同的组分。天然气凝析液用途广泛,可用于所有经济领域,可用作石油化工原料、烹饪燃料以及调和车用燃料。由于油价高企,天然气价格也有所上升,因此开采富含天然气的液体能源就有了动力。

## 🔲 课文

天然气凝析液来源于天然气,天然气是埋藏于地下的天然气态烃类混合物,可通过专门的驱替井开采。世界上不同地区天然气混合物的组分各不相同。其主要成分是甲烷,通常占其组分的 80% ~ 95%。其他成分还包括乙烷、丙烷、丁烷和其他烃类化合物,含量各不相同。

天然气凝析液是在天然气加工厂和炼油厂经冷却和蒸馏制成的,一般视为石油和天然气行业的"副产品"。天然气加工厂提取天然气凝析液是为了获取利润以及/或确保天然气符合管道输送标准。

天然气凝析液是随原油和天然气一起采出井的湿气、伴生气和伴生液。目前,天然气凝析液对于干气(甲烷)生产者十分重要,甚至可以说是至关重要,因为分离出的凝析液可以像乙烷、丙烷、丁烷、凝析油等产品一样高价出售。

鉴于天然气价格低迷,美国生产商不再钻所谓只产气的"干气"井,而是转向钻"湿气井",湿气井可产出气液混合体,其中液体更贵重,适用于石油化工业。

对非常规天然气开采有利的条件,同样有利于增加从富含液体的页岩气和轻致密油中提取的天然气凝析液(NGLs)的产量。轻致密油在许多方面和页岩气具有相似性,这种相似性既体现在轻致密油的来源(未离开过烃源岩(页岩),或至少未走远),也表现在其开发所需的生产技术上。在美国,轻致密油与非常规天然气大都产自相同的盆地,而且油价高,气价低,因此开采更多的液态烃具有很强的经济驱动力。

美国页岩气凝析液产量迅速增长,预计到 2020 年将新增高达 $100 \times 10^4$ bbl/d 的产能。天然气凝析液产量增加为石化行业带来了新的机遇,但想要实现效益最大化,还要克服输油管道的瓶颈障碍,并需要新增分馏、储存设备。从现在到 2020 年期间全球页岩气凝析液和轻致密油增加的产量(主要是北美),几乎将占据此期间石油供给增长量的一半。

然而 2020 年前,北美之外的页岩气凝析液和轻致密油的产量对全球石油产量增长的作用微乎其微,因为还要做许多评估工作。但阿根廷乌肯盆地很有发展前景,阿根廷石油公司(YPF)声称乌肯盆地有 $7 \times 10^8$ bbl 潜能,同时延伸到墨西哥的鹰福特页岩也是人们关注的一个焦点。据预测,即便 2020 年以后,北美以外的轻致密油产量仍然很小,鉴于目前资源勘探尚无取得长足发展,所以也许会有上升的可能。

然而,应该注意的是,从目前掌握的资料来看,轻致密油资源没有页岩气资源多:按照美国 2010 年国内天然气需求量来估计,美国页岩气资源至少能用 5 年;而已探明的轻致密油资源最多也只能满足国内 4 年的石油需求。这就是为什么我们预测美国轻致密油产量会在 21 世纪 20 年代达到顶峰。

凝析液含量是决定页岩气生产经济可行性的一个重要因素,因为天然气凝析液易于运至国际市场;而天然气一般只在当地市场交易,由于政策和基础设施等原因其价格无法与国际价格接轨。然而,新发现的页岩只有经过钻探和测试之后,才能确定其凝析液含量有多少。

未来美国液态天然气(天然气凝析液)增长量将取决于页岩层水力压裂完井的能力。但有关这一技术的争议越来越多。水力压裂即高压注入水、砂和化学品混合物,冲破并保持致密岩层张开,使原油、天然气和天然气凝析液流到地面。但环保组织认为这个开采过程消耗的水量大,会对地下水质构成威胁,并污染空气,甚至引发轻微地震。美国环境保护局(EPA)目前正在研究水力压裂对饮用水的影响,并计划制定国家级钻井废水处理标准。在过去的两年中,宾夕法尼亚州和得克萨斯州已经通过了关于压裂液成分公开的法律。

今年第一季度的天然气凝析液产量创纪录,约占美国石油总产量30%。页岩油的大量开采使北美天然气凝析液的价格大幅下滑,危及到天然气生产商的利益。天然气凝析液为许多页岩气生产商带来利益;但这有时也会降低页岩气的产量。在美国市场上一英热石油的价格是一英热天然气的20多倍。随着页岩气产量的增长,天然气凝析液可能会导致乙烷价格下降。这对顾客是好事,而对生产商却是坏事。乙烷从天然气中脱除后,可做乙烯原料,用于制造塑料。

页岩气凝析液的利润对于页岩气生产商来说十分重要,尤其是天然气价格较低时,所以乙烷价格下降可能会迫使一些生产商停止作业,直至乙烷供应量急剧下降,甚至不能满足需求时,其价格才会再次上涨。

通过下列数字可以看出天然气凝析液利润对页岩气生产商有多么重要:

- 页岩气井一般每一千立方英尺的产量达到5~6美元时才会有盈利。
- 去年甲烷售价基本上一直是4美元。
- 但凝析液附加值使页岩气井的全部产品价值约达7~8美元/$10^3 ft^3$。

许多专家曾讨论过为天然气及其凝析液建立一个交易市场,确保有买家愿意以适当的价格购买页岩气及其他非常规天然气,如致密砂岩气和北极气等。

## Key to Exercises

Ⅰ. 1. The chief component of NGL is methane, which usually makes up from 80% to 95% of its composition.

2. Natural Gas Liquids are the raw, associated gases and liquids that come up along with oil and natural gas from the well.

3. Hydraulic fracturing of wells involves injecting a high-pressure mix of water, sand and chemicals to help break and hold open dense rock formations, allowing crude, gas and NGLs to flow to the surface.

4. It may threaten underground water quality, pollutes the air and even causes minor earthquakes.

5. On the basis of current knowledge, light tight oil resources are expected

to be of less consequence than shale gas resources; whereas the estimated shale gas resources in theUnited States represent at least 5 years of 2010 domestic gas demand, the known light tight oil resources make up no more than four years of domestic oil demand.

Ⅱ. 1. D　2. A　3. B　4. C　5. D　6. C　7. B　8. B　9. A　10. A

Ⅲ. 1. B　2. D　3. C　4. C　5. B

Ⅳ. 1. 今年第一季度的天然气凝析液产量创纪录,约占美国石油总产量30%。

2. 对非常规天然气开采有利的条件,同样有利于增加从富含液体的页岩气和轻致密油中提取的天然气凝析液(NGLs)的产量。

3. 从现在到2020年期间全球页岩气凝析液和轻致密油增加的产量(主要是北美),几乎将占据此期间石油供给增长量的一半。

4. 页岩油的大量开采使北美天然气凝析液的价格大幅下滑,危及到天然气生产商的利益。

5. 随着页岩气产量的增长,天然气凝析液可能会导致乙烷价格下降。

Ⅴ. 1. 天然气凝析液是在天然气加工厂和炼油厂经冷却和蒸馏制成的,一般视为石油和天然气行业的"副产品"。天然气加工厂提取天然气凝析液是为了获取利润以及/或确保天然气符合管道输送标准。

2. 凝析液含量是决定页岩气生产经济可行性的一个重要因素,因为天然气凝析液易于运至国际市场;而天然气一般只在当地市场交易,由于政策和基础设施等原因其价格无法与国际价格接轨。然而,新发现的页岩只有经过钻探和测试之后,才能确定其凝析液含量有多少。

# 第2章 油气产业

## 2.1 油气产业结构

### 🔲 导语

非常规油气行业主要也分为三个部分：上游、中游和下游。上游部门主要从事非常规油气开发，不过开发技术和常规油气开发有所不同，而中下游生产与常规油气相同。因此，本章重点关注上游部分。上游部门寻求获得潜在油气田的开发权，以及进行和非常规碳氢化合物开采和生产相关的业务，通常有两类公司：天然气勘探生产公司和油田服务公司。因此，非常规油气勘探开发产业趋势和油田服务产业趋势也将成为我们关注的重点。

### 🔲 课文

石油和天然气行业主要可分为三个部门：上游、中游和下游。

• 上游公司业务包括油气勘探、开发和石油或天然气生产，也称为勘探开发公司。

• 中游业务往往是把比较偏远的产油区和消费区连接起来，具体包括油气加工、储存、销售和运输。

• 下游产业包括炼油厂、化工厂、油气配送公司，以及零售业务。

这些公司为消费者提供几千种石油产品，如汽油、柴油、天然气、燃料油（加热油）、柏油、润滑油、塑料、肥料、防冻剂、农药和丙烷等。

上游部门主要从事页岩气开发，而中下游业务与常规天然气相同。上游公司寻求获得油气田开发权，可进一步分为两类：天然气勘探开发公司和油田服务公司。

天然气勘探开发公司寻找潜在油气田，然后钻采地下岩层中的油气。公司的收入大部分源于出售井口油气。在美国，有两类油气勘探开发公司生产并销售天然气给管道运输公司和其他中游公司，分别是国际石油公司和独立石油公司。

国际石油公司是非国有的、公开交易的上市能源公司如：英国石油公司（BP）、埃克森美孚和荷兰皇家壳牌。这些公司业务常常延伸至中下游领域，上中下游一体化。

独立石油公司是非一体化的,相比于 BP 或埃克森美孚等大公司要小得多,其收益主要来源于井口气生产。这些公司业务极少涉及中下游,即使有的话,也仅限于运输和销售自产天然气。独立公司的规模大小不等,有市值数十亿美元并公开上市交易的大公司,也有小型独资企业。美国从事页岩气开发的独立天然气公司有切萨皮克能源公司、西南能源公司和油鹰能源公司等。

在美国之外的其他许多国家,国家石油公司也生产天然气。这些公司是国有性质,全部或部分归政府所有。国家石油公司掌控着大部分的世界油气储量。俄罗斯国家天然气工业有限公司、巴西国家石油公司、中国石油是世界上三个典型的国家石油公司。

最后,油田服务公司是服务于上游产业的公司,一般不自己生产石油赚取收益,而是通过为油气勘探开发作业提供技术、咨询、项目管理、油气井维护以及信息解决方案等业务获得收入。油服业务包括:各种科技服务,例如地震资料评估与分析、钻井设备供应与维护、工程服务以及钻头及压裂液供应。油服公司大小不等,既有提供专门产品和服务的小型独资公司,也有市值数十亿,提供一体化服务的大公司。世界上最大的三家油服公司是:斯伦贝谢、哈里伯顿和贝克休斯。

**上游油气公司职位**

上游油气公司的工作职位多种多样。天然气勘探开发公司和油服公司需要数百人从事不同层次的专业、技术和管理工作。勘探开发公司专业性职位占四分之一,80% 与建筑、工程和物理学相关。典型职业包括:工程师、地质专家、地球物理专家、化学及环境专家,绘图员、制图师和技术支持等。大部分职位都要求达到学士学位,有的可能要求更高。该行业技工一般需要 2~4 年的学历。具体职责如下:

● 地球物理专家要用重力勘探、磁法勘探、电法勘探和地震勘探等方法确定油气藏的潜在位置。

● 地质专家要勘探评估气藏的潜在产能。

● 地质信息专家创建并维护有关油井位置、管线、压缩机站、道路以及其他设施的信息库。

● 环境工程师负责制定并执行环保方案,对土壤和水质进行测试分析以确保符合环境法律法规。

● 油田工程师监管钻井、完井和修井作业。

地面工程和开采等非管理层职能是勘探开发公司第三大类工作,占其总职位的 16%。这些职位大多数和油气开采有关。直接与油气开采相关的职位有：负责生产的工头、油井工人和压缩机操作工。勘探开发公司还需要雇用挖掘机工人和伐木工人清理施工地点、修路、建井位、安置贮水池、建压缩机站、接好油气输送管线和电缆等。

油服公司和天然气开采公司招聘的人员在有些方面是相似的。然而，这两类公司招聘人员的组合是有区别的。在勘探开发公司，专业性职位占大多数；而在油服公司，生产性岗位占大多数。实际上，油服公司57%以上的岗位是井场劳务岗。

两种钻井工作就占油服公司人员的20%，即甲板工和旋转钻井工。其中甲板工所占比例最高，为油服公司总职位的15%。用特殊设备给油气井增产并清除油气井障碍物的工人占美国钻井服务部门近10%的比例。许多油服公司还雇用研发工程师以及产品项目经理来设计新产品和开发新技术，此外还雇用各种行业咨询及工程管理人员。

**上游油气公司就业前景**

2015年至2020年，就业率预计将增加3.6%，就业人数约达52.8万人。SBI Energy关于工业就业率增长情况的预测和美国能源信息署发布的2010年度能源展望报告相吻合。美国能源信息署估计美国油气产量会随着消费需求增加而上升，并预测原油价格会从2010年的约78美元/bbl上涨到2020年108美元/bbl以上；未来五年将新增63000个工作岗位，未来十年将增加82000个工作岗位。

另外，未来几年上游油气行业将面临大量的人员流失，据业内人员预测，到2019年将有30%～46%人员退休。由于油价上涨、钻井数量和开采量增加，而同时又有大量从业人员退休，所以很多新的工作岗位将出现，在未来的5～10年，将有大量的就业机会。对合格的专业人员和技术能力强、工作经验丰富的采油工人的需求量最大。

### 🔲 Key to Exercises

Ⅰ. 1. Upstream companies are involved in the exploration, development, and production of crude or natural gas, also called exploration and production, or E&P.

2. In the United States, two types of E&P companies produce and sell gas to pipelines and other midstream participants: international and independent oil companies.

3. These companies typically do not derive revenues from their own petroleum production, but instead provide technology, consulting, project management,

well services and maintenance, and information solutions to the oil and gas exploration and production industry.

4. Non-management functions related to construction and extraction are the third largest category of jobs in E&P companies.

5. In E&P companies, professional occupations account for the largest number of jobs. In contrast, production occupations account for the largest number of jobs in OFS companies.

Ⅱ. 1. A  2. C  3. D  4. B  5. A  6. C  7. B  8. D  9. D  10. C

Ⅲ. 1. B  2. A  3. C  4. D  5. B

Ⅳ. 1. 上游部门主要从事页岩气开发,而中下游业务与常规天然气相同。

2. 勘探开发公司的收入大部分源于出售井口油气。

3. 国家石油公司是国有性质,全部或部分归政府所有。

4. 地面工程和开采等非管理层职能是勘探开发公司第三大类工作,占其总职位的 16%。

5. SBI Energy 关于工业就业率增长情况的预测和美国能源信息署发布的 2010 年度能源展望报告相吻合。

Ⅴ. 1. 上游油气公司的工作职位多种多样。天然气勘探开发公司和油服公司需要数百人从事不同层次的专业、技术和管理工作。勘探开发公司专业性职位占四分之一,80% 与建筑、工程和物理科学相关。

2. 独立石油公司是非一体化的,相比于 BP 或埃克森美孚等大公司要小得多,其收益主要来源于井口气生产。这些公司业务极少涉及中下游,即使有的话,也仅限于运输和销售自产天然气。独立公司的规模大小不等,有市值数十亿美元并公开上市交易的大公司,也有小型独资企业。

# 2.2 页岩气就业前景

## 导语

以页岩气为主的非常规油气资源潜力巨大,已经在全球能源结构中扮演着重要的角色,是未来油气产业发展的重点。然而,页岩气生产需要进行复杂的水平钻井和多级水力压裂,因此,钻井和完井过程比常规天然气开发需要更多的劳动力。而且,页岩开发对经济发展和就业的影响远不只是给上游开发所创造的直接就业岗位。上游页岩气开发活动,通过间接消费对经济和就业产生了一系列影响。因此,页岩气开发将给总体就业形势带来美好前景。

## 课文

石油行业平均退休年龄为 55 岁,大量工作人员即将退休造成雇员青黄不接是这个行业现今面临的关键战略挑战。在 2007 年发布的一项研究中,州际石油天然气契约委员会估计,美国油气行业工人的平均年龄为 48 岁,其中一半为 50~60 岁。鉴于美国目前失业率较高,招聘新职员来补充退休人员还没问题。然而,退休的是经验丰富的工人,填补这一知识断层却很难,因为在石油行业,业务达到基本熟练程度需要长达三年的现场工作经验。

为此,在过去五年内,上游石油公司已采取各种措施吸引人才,应对众所周知的队伍大调整。行业组织已经成功说服联邦政府为地球科学专业的研究生和本科生增加实习机会;各州也提供资金培训矿区工人、油井服务工以及安全工程师等其他油田技术人员。同时行业协会加大公关力度,提升行业形象。

勘探开发和油服公司加强了与中学和大学的联系,为学校提供行业指导,提供奖学金和助学金,以支持在职人员的继续教育。公司还鼓励接近退休的员工继续留在公司,并以薪酬激励机制吸引新人才。据报道,2010 年石油工程毕业生起薪平均约达 75000 美元,使得石油工程成为年度最赚钱的本科专业。所有这些发展均表明,未来几年石油行业的就业前景更加光明。

**美国页岩气行业就业情况**

2010 年 7 月,美国实际工作的旋转转机共有 1585 个,其中 62% 是用来开采天然气的。SBI Energy 估计有三分之二的钻机是用于开发页岩气的,表明当时美国石油 40% 以上的钻井工作是为了开采页岩气。钻进活动是石油行业就业情况的晴雨表。从钻机数不仅能看出当前井场规模和钻井作业的就业率,也能看出后续钻井生产的就业率。

由于页岩气生产需要进行复杂的水平钻井和多级水力压裂,因此,钻井和完井过程比常规天然气开发需要更多的劳动力。事实上,马塞勒斯页岩开发教育和培训中心(MSETC)称,在马塞勒斯页岩气生产中,钻单井所需的直接工作人员超过 410 人,就是说每口井一年可以创造 11.53 个全职工作。然而,在完成钻井后,页岩气生产中每口井仅需 0.17 个全职工作人员,和常规天然气开采差不多。基于这些业务差异,以及 2010 年中期页岩气开发钻机比例,SBI Energy 估计石油行业上游产业的员工有 24%(107,000人)从事页岩气开发。

**美国页岩气行业就业情况预测**

美国能源信息署在其 2010 年度能源展望中预测,美国页岩气产量在

2015—2020 年将继续每年增长 3.2%，达到每年 $4.51 \times 10^{12} \text{ft}^3$ 的产量，即 $778 \times 10^6 \text{bbl}$ 油当量产量。相比之下，预计非页岩天然气产量在 2015 年后的 5 年里产量保持稳定。与此同时，原油产量 2015—2020 年会以每年 1.2% 的速率增长。

美国页岩气产量的增长需要持续钻进活动的支持，而这将促进就业。SBI Energy 预计 2015—2020 年员工数将会以每年 2.4% 的速度增长，2020 年将有 15.6 万名员工。

大量工作岗位需要具备以下条件：专业教育背景、天然气行业特殊培训、证书以及油田现场工作经验等。然而，页岩气开发也需要雇用大量传统性岗位工作人员，其中包括管理人员、商业运营和金融财会人员、信息技术员以及法律专家等。

### 美国页岩气开发形成的间接工作岗位

页岩开发对经济发展和就业的影响远不只是给上游开发所创造的直接就业岗位。上游页岩气开发活动，通过间接消费对经济和就业产生了一系列影响。页岩气开发的直接支出费用将滚动用于在其他经济领域。

例如，钻井之前须获得矿场开发权，这些工作和支出就创造了许多就业和经济机会。矿场租赁工作需要地质学家、调查员、丈量员、律师、文书、会计以及许多其他起草和敲定合约的人，而这些将会刺激许多非油气行业的发展。而且，通过矿产权租赁，数百万美国人拥有产区权益，每年总收益达几十亿美元。随着个人财富的增长，这些钱会通过各种渠道流入市场，进一步促进当地就业机会和经济发展。

2009 年宾州州立大学调查了前一年宾州马塞勒斯页岩气开发对当地经济的总体影响。该调查表明，矿区仅仅提供了 2100 个开采方面的直接工作岗位（油气行业占其中的一部分），但页岩气开发商的直接支出导致了间接的经济增长，创造了 27000 多个其他各行业的工作岗位。其中产生间接影响最大，且最明显的行业是运输和仓储行业，创造了约 4000 个岗位；建筑行业创造了将近 3800 个岗位；健康和社会服务也增加了 3500 个工作岗位。页岩气收益流入制造业、零售业、财务和保险以及房地产业，并为每个行业创造了 800 多个工作岗位。而对海内斯维尔马奈特和费耶特维尔页岩气开发进行的调查也得出了相似的结果：页岩气开发对当地的总就业率有非常显著的经济倍增效应。

## Key to Exercises

Ⅰ. 1. Industry organizations have successfully lobbied the federal government to increase internships for geosciences graduates and undergrads. States have provided funding for training programs for lease operators, well service crews, and other field technical resources such as safety engineers. In the meantime, industry associations have conducted massive public relations campaigns designed to rehabilitate the industry's poor image.

2. Petroleum engineering is the most lucrative undergraduate degree.

3. Drilling and well completion processes for shale gas wells are significantly more labor intensive than those for conventional gas as they leverage complex horizontal drilling and multi-stage hydraulic fracturing processes.

4. The jobs require various combinations of specific educational credentials, gas industry-specific training, certification, and field experience.

5. Shale gas development has a significant multiplier effect on the level of overall employment generation.

Ⅱ. 1. B　2. A　3. A　4. C　5. D　6. A　7. C　8. D　9. B　10. B

Ⅲ. 1. D　2. C　3. C　4. D　5. A

Ⅳ. 1. 然而,退休的是经验丰富的工人,填补这一知识断层却很难,因为在石油行业,业务达到基本熟练程度需要长达三年的现场工作经验。

2. 据报道,2010 年石油工程毕业生起薪平均约达 75000 美元,使得石油工程成为年度最赚钱的本科专业。

3. 然而,在完成钻井后,页岩气生产中每口井仅需 0.17 个全职工作人员,和常规天然气开采差不多。

4. 美国页岩气产量的增长需要持续钻进活动的支持,而这将促进就业。

5. 随着个人财富的增长,这些钱会通过各种渠道流入市场,进一步促进当地就业机会和经济发展。

Ⅴ. 1. 勘探开发和油服公司加强了与中学和大学的联系,为学校提供行业指导,提供奖学金和助学金,以支持在职人员的继续教育。公司还鼓励接近退休的员工继续留在公司,并以薪酬激励机制吸引新人才。

2. 页岩开发对经济发展和就业的影响远不只是给上游开发所创造的直接就业岗位。上游页岩气开发活动,通过间接消费对经济和就业产生了一系列影响。页岩气开发的直接支出费用将滚动用于在其他经济领域。

# 2.3 勘探开发产业趋势

## 导语

由于石油和天然气价格高企,水平钻井和水力压裂技术也均有所进步,勘探开发公司能通过页岩气开采盈利了,并寻求进一步加大页岩气开发规模。北美页岩气生产过去主要是由独立勘探开发公司承担;由于全球常规天然气和海上高质量油气资源越来越有限,开发成本也越来越高,国际石油公司也开始转向页岩气开发,以增加油气储量。实际上,埃克森美孚产品组成中大约75%为非常规油气资源。因此,可预见非常规油气勘探开发产业趋势前景辉煌。

## 课文

近年来,页岩气开发曙光初现,带来诸多长期发展机遇,吸引了许多新的市场参与者投身页岩气行业。由此产生了双重影响,一是改变了能源竞争格局,二是使页岩气从区域性产品开始转变为全球性产品。北美页岩气开发取得成功,引起了众多石油公司的关注,国际石油公司力图增加油气储备,外国国家石油公司力求获得非常规油气开采的专业技术。

直到最近,页岩气勘探和生产活动基本上都集中在美国和加拿大,不过加拿大参与程度比较有限。页岩气生产是由独立勘探开发公司主宰的。而国际石油公司主要从事海上和国外常规天然气以及液化天然气的生产。由于天然气价格高企,水平钻井和水力压裂技术也均有所改进,独立勘探开发公还是可以实现页岩气盈利的,并逐步积累大量页岩资产。

由于全球常规天然气和海上高质量油气资源越来越有限,开发成本也越来越高,国际石油公司也开始转向页岩气开发,以增加油气储量。随后,天然气价格回落了较长一段时间,这在很大程度上要归因于美国页岩气开采取得的成功,一些中小型开发商开始寻求战略方法来削减债务。

2008～2009年,在美国页岩气开发领域,独立石油公司与能源巨头之间的收购兼并活动日益频繁。例如,美国BP石油与切萨皮克能源公司合资,一同收购了伍德福德和费耶特维尔页岩气公司。随后,埃克森美孚斥资$410×10^8$美元收购了克洛斯提柏能源公司。此后,能源巨头与独立勘探生产公司之间的兼并收购活动持续不断,其中包括:道达尔与切萨皮克能源公司合资$22.5×10^8$美元,获得了切萨皮克能源公司美国页岩气资产股权,加快其开拓天然气市场的步伐;2010年5月荷兰皇家壳牌公司以$47×10^8$美元现金收购东方资源公司。

鉴于这种趋势以及许多令人羡慕的页岩气生产商在市场上已初显成就,可以预计能源巨头将继续从美国购买页岩气资源;这样在未来几年内,页岩气生产商的市场竞争格局将会有大幅改变。

### 埃克森美孚

#### 公司概述

埃克森美孚石油公司是世界上最大的油气公司,总部设在德克萨斯埃文市。公司的主要业务是勘探和开发原油及天然气,也包括石油产品加工和运输,以及原油、天然气和石油产品销售。埃克森美孚是世界上最大的石油炼制、化工产品制造及销售公司。尽管现在的埃克森美孚公司于 1999 年 9 月 30 号由埃克森和通用合并形成,但该公司的历史却可以追溯到 1870 年成立的标准石油公司。

2010 年 6 月,埃克森美孚公司决定以 $410 \times 10^8$ 美元收购克洛斯提柏石油公司。该公司是一家专门从事非常规天然气勘探和开发的公司,总部位于美国。2009 年 12 月底,根据克洛斯伯提能源公司报告,公司名下已探明的储量相当于 $148 \times 10^8 \mathrm{ft}^3$ 当量的天然气资源,包括页岩气、致密地层气、煤层气和页岩油。其中大部分储量分布在美国的三个页岩地层中:马塞勒斯、海内斯维尔和巴奈特。作为埃克森美孚的子公司,该公司命名为克洛斯提柏能源股份有限公司,致力于全球非常规油气资源开发。这次兼并对于埃克森美孚投资页岩气资产具有非常重要的意义,埃克森美孚也将借此扩大其在美国、加拿大、德国、波兰和阿根廷的份额。

#### 产品组成

2009 年年底,埃克森美孚已探明油气储量近 $230 \times 10^8 \mathrm{bbl}$ 石油当量。将近百分之五十的已探明储量是天然气资源,其余部分为凝析油、天然气凝析液、沥青以及合成油。该公司产品组成中大约 75% 为非常规油气资源。而且,兼并克洛斯提柏能源公司将进一步加大该公司在美国的非常规资源,尤其是页岩气的开发。

埃克森美孚一直在积极扩大非常规油气资源的开发,以解决近几年常规天然气日渐耗尽的问题。该公司致力于开发科罗拉多皮尔申盆地的致密气已经很多年,并称其在全球拥有大约 $700 \times 10^8 \mathrm{ft}^3$ 的非常规天然气资源。此外,并购克洛斯提柏能源又使其获得 $450 \times 10^8 \mathrm{ft}^3$ 以页岩气为主的非常规油气资源。

### 英国石油

#### 公司背景

BP 业务包括:石油天然气勘探开发、加工和运输,天然气、液化天然气和天然气凝析液的市场营销,以及原油和石化产品的炼制、营销和运输。BP 的替代能源业务在低碳能源方面,包括太阳能、风能、生物燃料、碳捕获与封存。BP 的业务遍及全球 80 多个国家,目前正在 30 个国家积极开展石油天然气开发和生产活动。

2008 年,BP 在美国页岩气市场上迈出了意义重大的一步,以 28.5 亿美元的现金收购了切萨皮克能源公司在伍德福德页岩气远景区(9 万英亩)和费耶特维尔页岩气远景区(1.35 万英亩)的租赁权。而且,BP 在海内斯维尔和伊戈尔福特的远景区也有一席之地。这四个区域的总资源潜力超过 $1 \times 10^{12} \text{ft}^3$。在国际上,BP 与俄罗斯合资的秋明英国石油公司正争取在乌克兰进行页岩气开发。

*产品组成*

BP(包括其子公司和股份实体)已探明储量超过 1800 万桶油当量。其中天然气占 40% 以上,剩下的为原油、凝析油以及天然气凝析液。在过去的五年里,公司已探明总储量增长了近 0.6%,其中原油储量增长了 2.4%,而已探明的天然气储量则下降了 1.7%。

截至 2009 年底,在 BP 总天然气资源中,常规资源占 62% 以上,其余为非常规资源,约占 BP 天然气产量的 10%。BP 指出页岩气资产已成为该公司在北美天然气业务中越来越重要的一部分。BP 在美国拥有包括海内斯维尔、费耶特维尔、伍德福德和伊格尔福特在内的超过 400acre 页岩气远景区的开发权。这四个北美页岩气远景区带潜在资源超过 $10 \times 10^{12} \text{ft}^3$。

## 🔲 Key to Exercises

Ⅰ.1. Over recent years, long-term opportunities presented by shale plays have attracted new participants to the shale gas industry.

2. These companies were able to realize profits from shale gas through the confluence of high prices for natural gas and the improvement of horizontal drilling and hydraulic fracturing technologies and further accumulate substantial shale asset portfolios.

3. Nearly 50 percent of ExxonMobil's proved reserves are in natural gas, with crudecondensates, NGLs, bitumen, and synthetic oil accounting for the balance. Approximately 75 percent of the company's portfolio is accounted for by unconventional oil and gas resources. As for BP, over 40 percent of the company's proved reserves are in natural gas; with crude, condensates, and NGLs comprising the balance. By the end of 2009, conventional sources accounted for over 62 percent of the company's total gas resources. Unconventional sources account for the balance and deliver about 10 percent of BP's natural gas production.

4. Their principal business is exploration and production of crude oil and

natural gas, as well as the manufacture of petroleum products and transportation and sale of crude oil, natural gas, and petroleum products.

5. BP's Alternative Energy business manages the company's activities in low-carbon energy, including solar, wind, biofuels, and carbon capture and sequestration.

Ⅱ.1. D  2. C  3. B  4. A  5. C  6. A  7. A  8. D  9. D  10. C

Ⅲ.1. C  2. C  3. B  4. A  5. B

Ⅳ.1. 埃克森美孚一直在积极加大非常规油气资源的开发,以解决近几年常规天然气日渐耗尽的问题。

2. 随后,天然气价格回落了较长一段时间,这在很大程度上要归因于美国页岩气开采取得的成功,一些中小型开发商开始寻求战略方法来削减债务。

3. 由于全球常规天然气和海上高质量油气资源越来越有限,开发成本也越来越高,国际石油公司也开始转向页岩气开发,以增加油气储量。

4. BP的替代能源业务在低碳能源方面,包括太阳能、风能、生物燃料、碳捕获与封存。

5. 将近百分之五十的已探明储量是天然气资源,其余部分为凝析油、天然气凝析液、沥青以及合成油。

Ⅴ.1. 直到最近,页岩气勘探和生产活动基本上都集中在美国和加拿大,不过加拿大参与程度比较有限。页岩气生产是由独立勘探开发公司主宰的;而国际石油公司主要从事海上和国外常规天然气以及液化天然气的生产。由于天然气价格高企,水平钻井和水力压裂技术也均有所改进,独立勘探开发公司还是可以实现页岩气盈利的,并逐步积累大量页岩资产。

2. 考虑到这一趋势和众多在市场上已初显成就而令人羡慕的页岩气生产商,可以预计能源巨头可能继续从美国购买页岩气资源,在未来几年,会大幅改变页岩气生产商的市场格局。

# 2.4  油田服务产业趋势

## 导语

油田服务公司是服务于上游产业的公司,通过为油气勘探开发作业提供技术、咨询、项目管理、油气井维护以及信息解决方案等业务获得收入。非常规油气资源潜能巨大,在全球能源结构中扮演着越来越重要的角色,并将成为未来油气产业发展的重点。因此,前景辉煌的非常规油气勘探开发产业也必将带来美好的油田服务产业趋势。本文将介绍世界两大油服公司,斯伦贝谢和哈里伯顿。

## 课文

油田服务公司为油气勘探开发商提供技术、咨询、项目管理、油气井服务和维护,给勘探开发行业提供信息解决方案。油田服务公司的客户包括独立勘探开发公司、国家石油公司和国际石油公司,这些公司需要油服公司提供基础设施、设备和技术来进行油气开采、储存和运输。

油田服务行业虽然系统性不强,但其架构大致可以分为四类:大型一体化全球性公司,如斯伦贝谢和哈里伯顿;小型专业化公司,提供技术、软件、工程服务或专用设备;提供廉价设备及普通技术或商品的公司;区域性公司。最近几年,该行业向着产品和服务多元化综合趋势发展,加强集产品和服务于一身的一体化服务能力,令客户享受更便捷的低成本服务。顺应这一发展趋势,油田服务公司一直在进行并购,旨在提高非常规天然气服务能力。

例如,2010 年 4 月,贝克休斯以 $55 \times 10^8$ 美元完成对 BJ 油田服务公司的收购,主要是因为该公司的压力泵送设备和用于页岩气及其他非常规气层的服务技术。而且,斯伦贝谢因其前合资方史密斯国际公司专长于页岩气开发,于 2010 年 8 月,以 $110 \times 10^8$ 美元全股收购该公司。史密斯国际公司是世界第二大钻头供应商,此前斯伦贝谢从未涉足过这一市场。这次收购使史密斯国际的计算机建模系统 I-Drill 也进入斯伦贝谢的资产组合当中,这一系统的功能是对钻井作业进行分析和预测。在过去的 24 个月里,哈里伯顿也一直忙于收购,收购那些能在技术上为开发非常规油气资源提供更多支持的公司,包括 Pinnacle Technologies,Evans Engineering,Permedia Research Group,Boots & Coots 等。

### 斯伦贝谢有限公司

#### 公司简介

斯伦贝谢公司成立于 1962 年,是世界上最大的油田服务公司。2009 年公司业务由两个部门经营:油田服务公司和西方物理信息勘探公司。油田服务公司为油气开发工程提供支持性产品和服务。西方奇科(Western Geco)是一家进行地表地震勘探数据采集和分析的公司。

2010 年八月,斯伦贝谢收购了史密斯国际股份有限公司——世界上第二大钻头生产商,并从此使斯伦贝谢进入了高速钻头行业,实现了产品和服务一体化经营,完全拥有 M-I SWACO 公司,M-I SWACO 公司以前和史密斯公司是合资公司。M-I SWACO 提供的钻井和完井液可使完井和修井作业对储层的伤害降至最低。

斯伦贝谢拥有 22 个先进科技项目研发中心,旨在提高油田工作效率、降低勘探和生产成本、提高产能、提升资产价值以及使储量开采最大化。为获取油气井井下数据,斯伦贝谢发明了电缆测井技术,并在试井、随钻测量、随钻测井、定向钻井服务、全电脑测井以及地球科学软件等方面成为领头羊。该公司业务遍布 80 个国家,在职员工有 10.5 万名;在巴黎、休斯顿和海牙都设有总部。

产品组成

斯伦贝谢以油田服务部门的各种产品服务于油气勘探和生产工程,包括电缆产品和服务、钻井和测量、完井及油井服务。此外,还提供软件、信息管理以及 IT 基础设施服务。

西方勘探公司提供综合性油藏绘图、监控和开发等服务,包括三维和四维及综合性地震勘探,用以描述油藏前景,进行油藏管理。而且,西方勘探公司拥有大规模地震数据处理中心,支持复杂数据处理项目。

**哈里伯顿公司**

公司概述

哈里伯顿是全球第二大油田服务公司,成立于 1919 年。为上游油气公司提供一系列完整的服务——包括油气勘探以及油井自始至终产能的最优化。哈里伯顿广泛的产品组合和一体化的服务能力,使其在油田服务市场上占据大量的份额,因为客户都力图签订一揽子服务合同,减少油田开发项目总成本。

哈里伯顿在 20 世纪 40 年代就成功实现了水力压裂的商业化发展,并在 20 世纪 80 年代巴涅特页岩开采中首次使用水平钻井和增产技术。此后,哈里伯顿为主要页岩盆地开发区都提供了各种各样的服务。哈里伯顿是为非常规油气开发公司提供技术的主要供应商,最近几年发展势头大增。哈里伯顿最近收购的公司包括:

- 尖峰科技,提供实时测斜仪、微地震绘图和油藏动态监测服务。
- 埃文斯工程股份有限公司,提供缆线和射孔产品。
- Permedia 研究小组,提供石油系统建模软件和服务。
- 布兹库兹公司,提供修井服务和压力控制产品。

产品组成

哈里伯顿包括两个业务部门:钻井评估和完井生产。钻井评估部门提供设备和以下服务:油藏评估,井位及稳定性模拟、测量以及优化等。产品和服务包括:油藏建模、钻头、钻井液和钻井、缆线和射孔、测试和海底业务、软件和资产方案以及项目管理服务。而完井和生产部门则提供固井、完井、增产等工具和服务。

## Key to Exercises

I . 1. Oilfield services ( OFS ) companies provide technology, consulting, project management, well services and maintenance, and information solutions to the oil and gas exploration and production industry.

2. Smith International is the world's second-largest provider of drill bits.

3. Schlumberger is the world's largest oilfield services company, and it maintained two business segments: Oilfield Services and Western Geco. The Oilfield Services segment provides products and services in support of oil and gas development projects. Western Geco is a surface seismic acquisition and processing company.

4. Schlumberger's technology development focuses on enhanced oilfield efficiency, reducing finding and production costs, productivity improvement, maximization of reserve recovery, and asset value improvement.

5. Halliburton maintains two business segments: Drilling and Evaluation and Completion and Production. Equipment and services to evaluate reservoirs and model, measure, and optimize well placement and stability are offered from the Drilling and Evaluation segment. Well cementing, completion, stimulation, and intervention tools and services are offered by the Completion and Production business unit.

II . 1. D  2. C  3. A  4. A  5. D  6. B  7. A  8. B  9. D  10. D

III . 1. D  2. D  3. A  4. B  5. C

IV . 1. 油田服务公司为油气勘探开发商提供技术、咨询、项目管理、油气井服务和维护,给勘探开发行业提供信息解决方案。

2. 顺应这一发展趋势,油田服务公司一直在进行并购,旨在提高非常规天然气服务能力。

3. 这次收购使史密斯国际的计算机建模系统 I-Drill 也进入斯伦贝谢的资产组合当中,这一系统的功能是对钻井作业进行分析和预测。

4. 油田服务公司为油气开发工程提供支持性产品和服务,西方奇科是一家进行地表地震勘探数据采集和分析的公司。

5. 该公司为上游油气公司提供一系列完整的服务——包括油气勘探以及油井自始至终产能的最优化。

V . 1. 斯伦贝谢拥有 22 个先进科技项目研发中心,旨在提高油田工作效率、降低勘探和生产成本、提高产能、提升资产价值以及使储量开采最大化。为获取油气井井下数据,斯伦贝谢发明了电缆测井技术,并在试井、随钻测量、随钻测井、定向钻井服务、全电脑测井以及地球科学软件等方面成为领头羊。

2. 哈里伯顿广泛的产品组合和一体化的服务能力,使其在油田服务市场占据大量份额,因为客户都力图签订一揽子服务合同,减少油田开发项目总成本。哈里伯顿在 20 世纪 40 年代就成功实现了水力压裂的商业化发展,并在 20 世纪 80 年代巴涅特页岩开采中首次使用水平钻井和增产技术。此后,哈里伯顿为主要页岩盆地开发区都提供了各种各样的服务。

# 第3章 非常规天然气开发

## 3.1 页岩气和致密气开发(1)

### 导语

页岩气和致密气都产自渗透率很低的油气藏,因此需要大量钻井,并进行水力压裂才能有经济效益,然而这对环境显然极具破坏性。最近在北美,为了减少井场建设对环境的影响和破坏,多底井受到了青睐。由于气体流速直接决定着油气井的经济效益,低流速就意味着得不偿失,没有投资回报,因此需要更多增产措施,提高天然气产量。

### 课文

页岩是富含黏土的地质岩层,通常由埋藏于海底或湖底几百万年的细粒沉积物形成。如果大量有机物随沉积物一起沉积,页岩中则会含有称为干酪根的有机固体物质。如果沉积过程中地层温度足够高,部分干酪根便会转化为石油或天然气(或两者的混合物)。

这一转化过程通常会使页岩储层压力增高,部分石油和天然气将向上运移至其他岩层,最终形成常规油气藏,因而页岩是常规油气藏的生油岩。但有一部分,有时是全部石油和天然气都会一直储存在其出生的页岩中,并最终形成页岩气藏或轻质致密油藏。

致密气藏的形成机理与常规气藏相同:由气源岩排出的天然气也恰巧运移至渗透率非常低的岩层。因此,致密气藏的商业化开采需要特殊技术。但致密气藏的采收率比页岩气藏高,因而单位表面积可采碳氢化合物的密度也较高。

#### 油井建设

钻井阶段是石油和天然气开发过程中最关键和最具破坏性的阶段,页岩气和致密气开采尤为如此,因为需要钻很多井。在陆地,钻机、配件、泥浆池及废料池等通常占地达1万平方米(井场)。到另一地点进行钻井作业时,单钻井设备就需要卡车来回运送100~200车次,而且钻井和完井过程中还需卡车运送物资。

选择井场时,不仅要考虑地下地质情况,还要考虑一系列其他因素,包括离人口稠密区和现有基础设施有多远,当地生态环境、水源及污水处置、气候的季节性限制或野生动植物保护等。最近在北美,为了减少井场建设对环境的影响和破坏,多底井受到了青睐。据行业资讯,2011 年,美国和加拿大新钻的页岩气和致密气井约 30% 是多底井。

钻井一旦开始,通常是 24h 连续作业,柴油机组不断产生噪音和烟雾,夜间需照明,卡车不断运送所需物资和设备。钻井作业或长或短,有时仅需短短几天,有时却耗时达好几个月,这主要取决于钻井深度和钻探的岩层类型。当钻头钻进岩层时,称之为"泥浆"的钻井液在井筒中循环流动,以便控制油井压力并将钻屑循环出来。润滑性"泥浆"由水或油等基液构成,同时混有盐和固体颗粒以增加密度,还含有多种化学添加剂。泥浆池有的是可移动的容器,有的是在地下挖的,并衬有防渗漏材料的露天池子。对泥浆池中的泥浆储量要进行监测以防渗漏或溢出。钻机一次使用的泥浆量很大,可能达数百吨。使用后的泥浆必须回收利用或进行安全处置。泥浆从每口井中携带出的岩屑量,根据钻井深度的不同,在 100 ~ 500t。这些岩屑,也需要适当处置,有利于环境。

钻井之后下钢质套管并注水泥可形成基本屏障,避免高压气体或液体从深地层逃逸到浅地层或含水层。这一屏障的设计必须得能承受住后期进行水力压裂时所产生的持续压力而不发生断裂。确保油井不渗漏最重要的几个设计方面是:钻井要严格符合规范(无随意转弯或腔洞);把套管置于井眼中心位置,然后再用水泥固定住,即在套管外围按均匀间隔放置扶正器,将套管下入井中,避免触碰井壁;此外,还要选择适合的水泥。水泥的配制既要考虑其泵入时的流体性质,以确保水泥能到达正确的位置,又要使其具备机械强度和灵活性,保证水泥环在固井后不变形。水泥的凝结时间也是关键因素,凝结时间过长可能会降低水泥的强度,同样,如果水泥还未完全泵入预设位置就凝固了,将很难再做调整。

**油井完井**

钻井结束后,横穿含气岩层的最后一段套管注完水泥后要进行射孔作业,以便岩层中的气体能够流入井筒。此时,油井中的压力会降低,由于压力差,碳氢化合物便可从周围岩层流入井筒。页岩气和致密气储集岩层渗透率非常低,因而气体流量也非常低。由于气体流速直接决定着油气井的经济效益,低流速就意味着入不敷出,没有投资回报。如果不采取进一步的增产措施,开采作业就无法带来经济效益。

# Key to Exercises

Ⅰ. 1. Shales are geological rock formations rich in clays, typically derived from fine sediments, deposited in fairly quiet environments at the bottom of seas or lakes, having then been buried over the course of millions of years.

2. The reason is that a larger number of wells are required. On land, a drilling rig, associated equipment and pits to store drilling fluids and waste typically occupy an area of 100 metres by 100 metres. Setting up drilling in a new location might involve between 100 and 200 truck movements to deliver all the equipment, while further truck movements will also be required to deliver supplies during drilling and completion of the well.

3. This "mud" consists of a base fluid, such as water or oil, mixed with salts and solid particles to increase its density and a variety of chemical additives.

4. A combination of steel casing and cement in the well provides an essential barrier to ensure that high-pressure gas or liquids from deeper down cannot escape into shallower rock formations or water aquifers.

5. Once the well has been drilled, the final casing cemented in place across the gas-bearing rock has to be perforated in order to establish communication between the rock and the well.

II. 1. A  2. A  3. B  4. D  5. A  6. C  7. D  8. A  9. A  10. D

III. 1. C  2. A  3. C  4. A  5. A

IV. 1. 这一转化过程通常会使页岩储层压力增高,部分石油和天然气将向上运移至其他岩层,最终形成常规油气藏。

2. 钻井之后下钢质套管并注水泥可形成基本屏障,避免高压气体或液体从深地层逃逸到浅地层或含水层。

3. 钻井阶段是石油和天然气开发过程中最关键和最具破坏性的阶段,页岩气和致密气开采尤为如此,因为需要钻很多井。

4. 钻井一旦开始,通常是 24 小时连续作业,柴油机组不断产生噪声和烟雾,夜间需照明,卡车不断运送所需物资和设备。

5. 水泥的配制既要考虑其泵入时的流体性质,以确保水泥能到达正确的位置,又要使其具备机械强度和灵活性,保证水泥环在固井后不变形。

V. 1. 如果大量有机物随沉积物一起沉降,页岩中则会含有称为干酪根的有机固体物质。如果沉积过程中地层温度足够高,部分干酪根便会转化为石油或天然气(或两者的混合物)。

2. 页岩气和致密气储集岩层渗透率非常低,因而气体流量也非常低。由于气体流速直接决定着油气井的经济效益,低流速就意味着入不敷出,没有投资回报。如果不采取进一步的增产措施,开采作业就无法带来经济效益。

# 3.2 页岩气和致密气开发(2)

## 导语

水力压裂和酸化处理是页岩气和致密气开发中提高气体流动性的两项重要技术。如果只用酸来溶蚀地层就是酸压。水力压裂技术初次试用是在1947年,而1949年就成功用于商业开发了。作为增产技术水力压裂通常在气井生命周期中只使用一次,就能大大增强流体的流动性和气井产能,但现在由于产能下降,多次压裂应用越来多了。然而,这些行为造成的环境威胁不容忽视,比如地下水污染、淡水枯竭、空气污染、噪声污染以及反排液和溢出流体对地表的污染等。

## 课文

### 酸化处理和水力压裂

近年来,为提高低渗透率油气藏流体的流动性,开发了多项增产技术。其中酸化处理技术出现得最早,如今仍广为使用,尤其适用于碳酸盐储层。该技术的原理是通过向气藏注入少量的强酸来溶解部分岩石矿物,进而提高井筒附近岩石的渗透率。水平层段长的井(也称水平井)可以使储集岩和井筒的接触面积大幅增加,提高开采效益。水力压裂技术兴起于20世纪40年代末,也常用于有效开发低渗透气藏。若岩石渗透率极低,如页岩气藏或轻质致密油藏,往往选择水平钻井加水力压裂的方法,才能实现商业化生产。水平井和水力压裂技术联合使用,是美国2005年页岩气产量猛增的主要原因。

水力压裂技术的原理是沿气井高压注入压裂液,打入地下很深的气井周围的目标岩石,使岩石形成几毫米宽的裂缝或裂隙。这些裂缝在井筒中可以延伸几十米,有时甚至长达几百米。然而,一旦压力不足,这些裂缝很容易再次闭合,难以持续改善碳氢化合物的流动性。为了保持裂缝处于张开状态,要将石英砂或陶瓷粒等小颗粒物添加到压裂液中,和压裂液一起泵入,作为支撑剂填充裂缝,也就是说撑开裂缝,让气体逸出进入井筒。

在许多情况下,需要按照规定间隔进行一段一段的系列压裂,即沿水平井筒每隔约100米压裂一次。这种多段压裂技术在美国页岩气和轻质致密油生产中发挥了关键性作用,且有希望在全世界范围内使用。一段水力压裂一般要注入几百立方米水,以及支撑剂和各种化学添加剂。页岩气井一般需要10~20段压裂,因此,再乘以10~20倍的水和固体量,每口井的总用水量可能会达到几千到两万立方米,支撑剂用量会达到 1000～4000t。多段高压压裂作业会对井产生反复应力,这就要求钻井设计和施工投入更多,以确保含气层与气井穿过的其他地层完全隔离。

水力压裂一旦完成,部分注入的流体会随产出流体一起沿井返排出来,但不会全部排出,一部分仍然会滞留在岩石中。返排通常会持续数天(一段压裂)到数周(多段压裂)。在此期间,排出的压裂液越来越少,而采出流体中的烃含量逐渐增加,直到井流基本上都是烃类物质。

这一过程最好进行所谓的"绿色完井"或"减少排放完井",这样可以从压裂液中分离出烃来出售,将残余返排压裂液收集起来,处理后再循环使用或废弃。然而,尽管按常规应该收集和处理加工返排流体,但是在返排初期收集并出售气体需要投资配备气体分离和加工设施,这不是都能做到的。鉴于此类情况,可以将气体排放到大气中(其主要成分是甲烷,含有小部分挥发性有机化合物),或点火处理(燃烧)烃或烃/水混合物。因此,该阶段排放或燃烧气体是促使页岩气和致密气生产中温室气体排放量比常规油气生产高的主要原因。

### 生产

一旦油井与处理设备连通,即进入主要生产阶段。油井产出的碳氢化合物和废液要设法处理。但是现在井场处理设备并不显眼:由阀门组成的"采油树"通常有一米高,位于井的顶部。采出的气体通过管道输送到处理中心,一个处理中心可以服务好几口油井。有时生产商会决定在生产井寿命后期重复进行水力压裂作业,即重复压裂。重复压裂常用于垂直井,水平井较少使用,美国进行重复压裂的水平页岩气井不到 10%。

油气井生命周期中生产阶段时间是最长的。常规油气井可持续生产 30 年或者更长。至于非常规油气开发,油气井的开采寿命预计与常规井差不多。但页岩气井通常初始产量高,随后产量急剧下降,然后在相当一段时间内产量很低。第一年产量通常会下降 50%～75%,大部分的可开采油气通常都是在之后的几年里才开采出来。

### 弃井

油井失去经济效益时应安全弃井,拆除各种设备,将井区土地恢复原始状态或适当投入新的生产。要注意长期预防井内流体进入含水层或地表。只有停产才会弃井,因此,有关规章制度应保证相关公司留有足够的财力和技术维护能力,在油田经济寿命结束后,还能保证弃井工作顺利完成,从长远角度维护井的完整性。

## 🔲 Key to Exercises

Ⅰ. 1. Acid treatment involves the injection of small amounts of strong acids into the reservoir to dissolve some of the rock minerals and enhance the permeability of the rock near the wellbore.

2. Multi-stage fracturing technique.

3. Because capturing and selling the gas during this initial flow-back phase requires investment in gas separation and processing facilities, which does not always take place.

4. For a conventional well, production might last 30 years or more.

5. When wells are safely abandoned, facilities should be dismantled and land returned to its natural state or put to new appropriate productive use. Long-term prevention of leaks to aquifers or to the surface is particularly important.

Ⅱ. 1. A  2. B  3. D  4. A  5. B  6. C  7. A  8. C  9. A  10. A

Ⅲ. 1. A  2. B  3. B  4. D  5. A  6. C  7. A  8. D  9. B  10. B

Ⅳ. 1. 返排通常会持续数天(一段压裂)到数周(多段压裂)。在此期间,排出的压裂液越来越少,而采出流体中的烃含量逐渐增加,直到井流基本上都是烃类物质。

2. 在许多情况下,需要按照规定间隔进行一段一段的系列压裂,即沿水平井筒每隔约100米压裂一次。

3. 有时生产商会决定在生产井寿命后期重复进行水力压裂作业,即重复压裂。

4. 这一过程最好进行所谓的"绿色完井"或"减少排放完井",这样可以从压裂液中分离出烃来出售,将残余反排压裂液收集起来,处理后再循环使用或废弃。

5. 油井失去经济效益时应安全弃井,拆除各种设备,将井区土地恢复原始状态或适当投入新的生产。

Ⅴ. 1. 近年来,为提高低渗透率油气藏流体的流动性,开发了多项增产技术。其中酸化处理技术出现的最早,如今仍广为使用,尤其适用于碳酸盐储层。该技术的原理是通过向气藏注入少量的强酸来溶解部分岩石矿物,进而提高井筒附近岩石的渗透率。

2. 水力压裂技术的原理是沿气井高压注入压裂液,打入地下很深的气井周围的目标岩石,使岩石形成几毫米宽的裂缝或裂隙。这些裂缝在井筒中可以延伸几十米,有时甚至长达几百米。然而,一旦压力不足,这些裂缝很容易再次闭合,难以持续改善碳氢化合物的流动性。为了保持裂缝处于张开状态,要将石英砂或陶瓷粒等小颗粒物添加到压裂液中,和压裂液一起泵入,作为支撑剂填充裂缝,也就是说撑开裂缝,让气体逸出进入井筒。

# 3.3 煤层气开采

## 导语

煤层气这个术语指的是吸附在固体煤上的甲烷,可以从煤层中开采出来。一些煤层富含气早已为人所知,因此,出于安全,人们从地面钻井到煤层,将甲烷释放出来后再进行采煤。20世纪70年代末美国联邦政府就力推将煤层气作为天然气资源开采,美国能源部资助许多非常规天然气资源的研究,包括煤层气。煤层气不但免于联邦政府的价格控制,还给予税收抵免。澳大利亚是1996年开始对煤层气进行商业化开采的。

## 课文

煤层气是指在煤层中的甲烷气体(天然气)。一些甲烷吸附在煤层中,在煤层孔隙表面形成一层吸附膜。煤层中张开的裂缝也可能含有游离气和游离水。有时煤层中含有大量甲烷,给采煤作业带来重大安全隐患。煤层中可能还有大量的二氧化碳。

煤层气同其他两种讨论过的主要非常规气既相似,也有差异,主要体现在开采方式、开采成本以及对环境的影响等方面。相同之处主要在于这些气藏的渗透率都很低,是影响开采技术和经济可行性的重要因素。实际上,煤层渗透率都与裂缝节理有关。裂缝是天然形成的,甲烷可通过这些裂缝在煤层中流动。与页岩气和致密气藏相似的是,煤层中各处的天然气含量不同,如果煤层厚度再不一样,将对潜在产量有重大影响。

在地表开采煤层气会破坏当地景观和自然环境,因为要在地面搭建钻台、修建道路、安装现场生产设备和气体处理及集输设施等。与页岩气和致密气开采一样,煤层气开采中的钻井数量比常规油气开采要多得多。因此,运输车辆和压缩机等带来的噪声污染、空气污染以及对当地生态系统的破坏就更严重。

煤层气与页岩气和致密气有着显著差异。煤层气埋藏较浅(主要限于目前所开采的),而页岩气和致密气通常埋藏较深。煤层中经常有水,因此必须排水才能使气体流入井中。另外,煤层气中的重质液态烃含量较少(天然气凝析液或凝析油),因此煤层气开采的商业价值很大程度上取决于甲烷气的价格;而页岩气生产中同时伴有大量的天然气凝析液,因此页岩气的总体开发效益主要取决于油价,如图3.1所示。

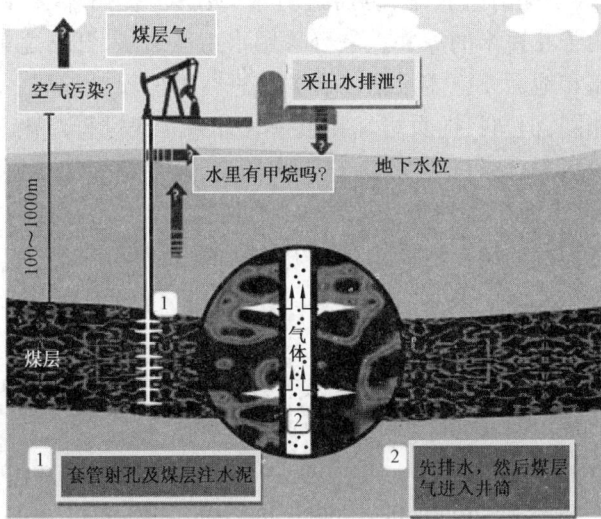

图 3.1　煤层气开采技术和可能造成的环境危害

过去25年来,用珩磨技术对煤层气进行商业化开采取得了重大进步,为大规模生产奠定了基础,北美率先对煤层气进行大规模开采,自20世纪90年代中期起澳大利亚也开始了煤层气的大规模开采。

直井或水平井都可用于煤层气开采,但水平井用得越来越多(尽管不如页岩气开采那样普遍)。一般情况下,如果煤层薄而埋藏又深,就更倾向于打水平井。虽然煤层气一般埋藏于地下800~1200米,但有时地表以下100米左右的浅层也会含有煤层气,因此,打一系列直井更经济,不用沿煤层打水平井。如果储层埋藏较浅,用水井设备钻井即可,比用常规油气钻机成本低多了。如果储层埋藏较深(400~1200米),直井和水平井都要用,已经开发出特殊定制的能控制井喷的小型钻机。

井钻好后,就要利用自然压力或机械泵将煤层中的水采出,这一过程叫作气井排水。随着排水过程的进行,地下压力下降,原来因水压而滞留的天然气从煤层的天然裂隙或节理中释放出来,加速流动。在地面,将天然气从水中分离出来,经压缩后注入天然气集输管道。

像页岩气一样,煤层气采收率要比常规气低得多;排完水后,产量会迅速达到峰值,之后,随着井内压力不断下降,产量就进入递减期。煤层气井的寿命一般是5~15年,排水完成后的1~6月是气井产量最高时期。大多数情况下,由于煤层的自然渗透率低,所以只有一小部分煤层的气体可以流入井中,页岩气、致密气开采同样如此。因此,煤层区需要大量钻井,如果钻直井,需要的数量就更多。

有时候,为了提高采收率,也需进行水力压裂,提高煤层渗透率。水力压裂一般应用于几百米的深井。是否采用水力压裂法,开钻之前就需确定,因为井的分布和地面设施需根据压裂方法进行专门设计部署。压裂方法同上面描述的相同,但与目前页岩气和致密气开采实践进行对比,煤层气开采通常采用单段压裂,即每口井只进行一次压裂,而不是多次压裂。由于井是分批钻的,所以水力压裂所需用水可从附近正在排水的井抽取。井中返排流体要泵入防渗漏的储存池或储存罐中,供后续处理或排放。

## Key to Exercises

Ⅰ. 1. "This" refers to that there are major variations in the concentration of gas from one area to another within the coal seams.

2. Above ground, coalbed methane production involves disruption to the landscape and local environment through the construction of drilling pads and access roads, and the installation of on-site production equipment, gas processing and transportation facilities.

3. Coalbed methane deposits can be located at shallow depths, whereas shale and tight gas are usually found further below the surface. Water is often present in the coalbed, which needs to be removed to allow the gas to flow to the well. In addition, coalbed methane contains very few heavier liquid hydro-carbons, which means the commercial viability of production depends heavily on the price at which the gas itself can be sold; in the case of shale gas produced together with large volumes of associated natural gas liquids, the price of oil plays a very important role in determining the overall profitability of the devel-opment project.

4. In most cases, the low natural permeability of the coal seam means that gas can flow into the well from only a small segment of the coal seam-a charac-teristic shared with shale and tight gas.

5. Because the well and surface facilities need to be designed accordingly.

Ⅱ. 1. A  2. A  3. B  4. C  5. B  6. D  7. C  8. A  9. D  10. B

Ⅲ. 1. A  2. B  3. C  4. D  5. B  6. A  7. B  8. C  9. D  10. A

Ⅳ. 1. 一些甲烷吸附在煤层中,在煤层孔隙表面形成一层吸附膜。

2. 有时煤层中含有大量甲烷,给采煤作业带来重大安全隐患。

3. 过去25年来,用珩磨技术对煤层气进行商业化开采取得了重大进步,为大规模生产奠定了基础,北美率先对煤层气进行大规模开采,自20世纪90年代中期起澳大利亚也开始了煤层气的大规模开采。

4. 在地面,将天然气从水中分离出来,经压缩后注入天然气集输管道。

5. 井中返排流体要泵入防渗漏的储存池或储存罐中,供后续处理或排放。

V.1. 在地表开采煤层气会破坏当地景观和自然环境,因为要在地面搭建钻台、修建道路、安装现场生产设备和气体处理及集输设施等。与页岩气和致密气开采一样,煤层气开采中的钻井数量比常规油气开采要多得多;因此,运输车辆和压缩机等带来的噪声污染、空气污染以及对当地生态系统的破坏就更严重。

2. 煤层气与页岩气和致密气有着显著差异。煤层气埋藏较浅,而页岩气和致密气通常埋藏较深。煤层中经常有水,因此必须排水才能使气体顺利流进井中。另外,煤层气中的重质液态烃含量较少,因此煤层气开采的商业价值很大程度上取决于甲烷气的价格;而页岩气生产中同时伴有大量的天然气凝析液,因此页岩气的总体开发效益主要取决于油价。

# 3.4 煤层气开采及对地下水的影响

## 导语

像页岩气一样,煤层气采收率要比常规气低得多。为了提高采收率,也需进行水力压裂,提高煤层渗透率。然而,由于煤层气产层比致密气和页岩气藏浅,所以压裂液进入含水层的风险更大。有时井流会通过套管渗漏到含水层,甲烷也可能会通过岩石渗漏。同页岩气和致密气生产一样,压裂作业之后的返排液需经处理才能排放。但有时污水未经充分处理就排入地下水了。

## 课文

煤层气开采对地下水以及煤层附近含水层的水供应和水质造成的影响令人堪忧。影响程度取决于地理位置及其他一些因素,其中最重要的因素有煤层原生水的总体积以及盆地水文地质情况;煤层气井的密度;泵水速度;煤层和含水层与周围水源的连通情况;以及含水层的补水速度和泵送所花时间等。

在美国,许多机构为了解生产过程,对煤层气开采区进行水监测。有文献记录,煤层气开采导致粉河盆地含水层枯竭,主要位于盆地的蒙大纳区域;停产后,煤层地下水 65% ~87% 已采出。然而,对浅层冲积扇和地下含水层的水位下降程度没有测量(最近澳大利亚昆士兰州立法要求对以上两项进行测量)。有证据表明,某些地区煤层气生产中气井排水所引起的地下水运动,使盐溶解量和其他矿物含量增多。

由于煤层气产层往往比致密气和页岩气藏浅,因而压裂液直接或间接通过裂隙(自然形成的裂隙或经压裂形成的裂隙)进入含水层的风险也更大。跟页岩气和致密气开采比起来,由于开采煤层气需要进行气井排水,所以气井附近含水层里剩下的水会少一些,在一定程度上减少了水污染的可能性。同页岩气和致密气生产一样,压裂作业之后的返排液需经处理才能排放。

第一大风险是地表溢流风险,严格密闭液相和固相流体可降低这类风险。意外事故在所难免,但制定良好的工作规程,加强对人员的培训和安装泄漏控制设备可一定范围内控制风险。尽量多使用管道运输液体,可以降低卡车运输风险。

第二大风险是套管渗漏,污染附近含水层。为了控制该风险,就要优化气井设计和气井建设,尤其是注水泥阶段,要确保密封到位,系统验证密封质量,并确保在气井的生产期内密封完好无损。对进行多段水力压裂的气井来说防渗漏更是重要了:反复进行高压泵送会对套管和水泥环产生反复应力,削弱套管和水泥环的强度。因此,选择适当强度的套管很重要。

第三大风险是产层油气穿过周围岩层泄露,这种现象在页岩气和致密气开采中不可能出现,因为这两种气藏的产层位于相关含水层以下一千至几千米深的地方,而且产层和含水层之间通常由一个或多个渗透性极差的岩层隔开。例如,在 Barnett 页岩区,地下饮用水源最深的达 350m,而页岩层则位于 2000 ~2300m 深的地层。然而,如果产层较浅或目标层上方含有浅层天然气,则同样会有危险。如果产层和含水层之间没有非渗透性岩层,或有深断裂带,那么断裂带就像导管一样可以使液体从产层深处流向地表(这种液流速度一般较慢,但可流动几十年),这种情况从理论上讲是有可能发生的。另外,水力压裂可能不仅仅压裂目标岩石层,也许还会压裂重要的岩层屏障,或联通断裂层。不过这种现象罕见,因为设计压裂方案时要避免这种情况的发生,但如果对局部地质情况了解不够的话,也不能完全排除这种可能性。

因此,在进行大规模开发之前,必须对当地地质状况进行预先研究,了解情况。事实上,甲烷气体可渗透至地表已久为人知(例如,在印尼爪哇的

Mrapen 村燃烧了几个世纪甲烷火焰,还有美国纽约州的"永恒之火瀑布"),人们就是依靠这些信号找到了地下油气藏,表明岩石密封并不总是完美无缺的。

另一方面,油气渗流以及由此导致的甲烷气侵入含水层都表明,不是所有的污染都与工业活动有关;也可能由自然地质或生物过程作用导致。第四大风险是将未充分处理的污水排入地下水,甚至地下深层。解决这一问题需要实施一系列监管政策,包括追踪并记录污水排放量、污水成分以及污水运输和处理方式。

## Key to Exercises

Ⅰ. 1. We show concerns over the impact ofcoalbed methane production on groundwater flows and the supply and purity of water in aquifers adjacent to the coal seams being exploited.

2. Because productive coal seams are often at shallower depths than tight or shale gas deposits in coalbed methane production.

3. When multi-stage hydraulic fracturing is performed, the repeated cycles of high pressure pumping can apply repeated stress to the casing and to the cement column, potentially weakening them. Therefore, selection of an appropriate strength of casing can prevent leakage into a shallow aquifer behind the well casing.

4. Because the shale gas or tight gas producing zone is one to several thousand metres below any relevant aquifers and this thickness of rock usually includes one or several very impermeable layers.

5. Itrequires a regulatory response, including appropriate tracking and documentation of waste water volumes and composition, how they are transported and disposed.

Ⅱ. 1. A  2. A  3. C  4. C  5. A  6. B  7. B  8. D  9. B  10. A

Ⅲ. 1. D  2. D  3. A  4. A  5. A  6. A  7. A  8. D  9. C  10. B

Ⅳ. 1. 煤层气开采对地下水以及煤层附近含水层的水供应和水质造成的影响令人堪忧。

2. 有证据表明,某些地区煤层气生产中气井排水所引起的地下水运动,使盐溶解量和其他矿物含量增多。

3. 同页岩气和致密气生产一样,压裂作业之后的返排液需经处理才能排放。

4. 如果产层和含水层之间没有非渗透性岩层,或有深断裂带,那么断裂带就像导管一样可以使液体从产层深处流向地表,这种情况从理论上讲是有可能发生的。

5. 因此,在进行大规模开发之前,必须对当地地质状况进行预先研究,了解情况。

V.1. 由于煤层气产层往往比致密气和页岩气藏浅,因而压裂液直接或间接通过裂隙进入含水层的风险也更大。跟页岩气和致密气开采比起来,由于开采煤层气需要进行气井排水,所以气井附近含水层里剩下的水会少一些,在一定程度上减少了水污染的可能性。同页岩气和致密气生产一样,压裂作业之后的返排液需经处理才能排放。

2. 水力压裂可能不仅仅压裂目标岩石层,也许还会压裂重要的岩层屏障,或联通断裂层。不过这种现象罕见,因为设计压裂方案时要避免这种情况的发生,但如果对局部地质情况了解不够的话,也不能完全排除这种可能性。

# 第4章 非常规油气开发技术

## 4.1 勘探

### 导语

在油气勘探中,地震勘探方法应用最为广泛。与其他任何地球物理方法相比,如电法勘探、磁法勘探和重力勘探,地震勘探能提供更为详细的地下地质结构形态和深度方面的资料。地震勘探可以用特殊的地震仪或检测器,记录由人工爆炸或非爆炸地震技术所产生的地面运动,探查地下沉积带的地质构造。在反射地震学中,地震资料处理的最终成果是类似于地质横剖面的地震剖面。在地震剖面上可以看出反射层,解释员可以利用地震剖面来绘制地下地质构造。

### 课文

勘探是油气资源开发过程中关键性的第一步。在地层深处开发页岩气,首先要精确估算其潜在储量的位置和质量,并确定气藏的地质和地球物理特性,然后再进行钻井、增产和开采等步骤,最终使页岩气生产的经济可行性和产量最大化。勘探开发公司大力投资勘探技术,以实现产量最大化,并将钻探无产能区域的风险和成本降到最低。

页岩气勘探一般是地质学家先勘查地表结构,确定气藏可能存在的区域,并从地质学角度识别可能的有天然气储层的区块,然后再用大量先进技术来精确绘制地层图。这些先进技术包括地震勘探、磁法勘探和重力勘探。

**二维反射地震学**

基础地震学是应用最为广泛且效果最好的技术,如图 4.1 所示。地震学研究的是地震波穿过地壳的运动,以及震波在不同类型地层中的相互作用。反射地震勘探法常用于绘制地下岩层结构图。

反射地震学是人工产生地震波,然后抓取震波在地下的反射,以便于多维呈现地下地质状况。过去用炸药产生地震波,而现在大多数公司采用非爆炸地震技术,即用载有特殊设备的轮式或者履带式车辆来制造系列震波。这些地震波在地层中传播并在不同地层表面反射。反射的地震波由放置地表或嵌入地表的敏感性检波器接收。检波器将接收到的数据随后传送给地

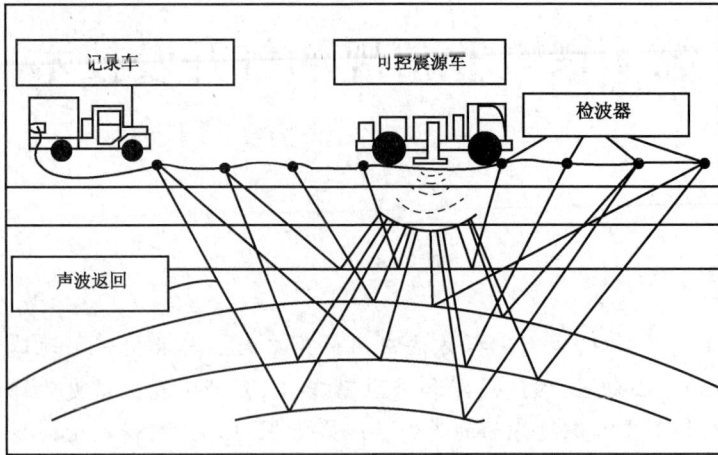

图 4.1　二维反射地震过程透视

震波记录车存储,留待地球物理学家和石油资源工程师对其作进一步解释。

　　直到 20 世纪 80 年代,反射地震勘探都是把震源和接受器排列在一条线上,然后一路推进设备完成勘探,最后生成地下横截面的二维地震波剖面。几十年来,二维地震勘探广泛用于油气工业来绘制常规油气资源图。然而,通过二维地震剖面确定高产井位的成功率仅为 50% 左右。

**三维地震成像**

　　随着勘探开发(E&P)公司勘探的常规和非常规油气资源距地面越来越远,埋藏也越来越深,这些资源的开发成本和风险也越来越高,因此,地下勘探的准确度就显得越来越重要。三位地震勘探与二维地震学的原理相同,只不过是用的地震检波检器更多,将其布置成网格状,从而生成颗粒更小的地下三维视图。从 20 世纪 90 年代起,三维地震技术的应用逐步增多,虽然三维勘探成本更高些,但精确定位高产区域的概率大幅增加,足以抵消其成本费用。

　　最近数据采集技术取得了进步,使三维探测技术的质量和经济效益均有所提高。陆地进行三维地震勘探时,检波器接收站和电缆布设开销庞大,严重限制了检波器的使用数量。高等电子学理论的应用使接收仪信号基本上实现数字化了,数据可以本地记录,也可以通过雷达或者光纤转至基站。数据处理技术的进步,使地震勘探中布设更多的检波器经济可行。此外,地震数据处理和成像技术的进步也提高了预测结果的精确性。实际上,使用三维地震勘探技术比用二维技术可能会使储层位置定位成功率提高 50% 。

　　这些进步以及过去二十年间计算机技术所取得的进步都使三维地震勘

探数据采集与处理更加经济可行。因此,大多数地震数据都是以三维地震勘探方式收集的,如图4.2所示。

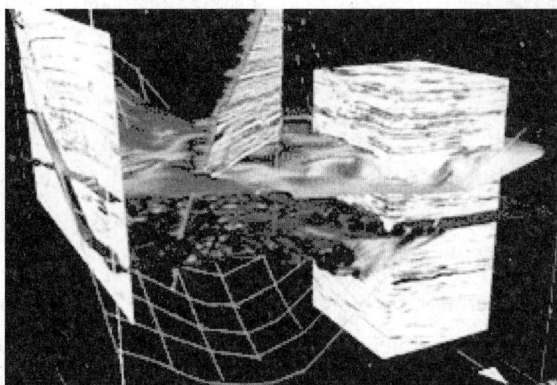

图4.2　高级三维地震图像

**四维地震成像**

最新技术发展是四维地震技术。四维地震技术在三维地震技术基础上增加了时间作为第四维,这样可以观察到地下地质特征在时间上的变化。因为时间是第四维,这项技术也叫作四维延时成像技术。本质上来说,四维地震信息是按时间顺序抓取的,是为了建立地下地层的动态模型,通过分析这些时间上的变化,可以更好地认识地下流体的流动性、粘性、温度及饱和度等特性。

利用四维勘探技术,勘探队可以更容易地确定天然气的前景,更有效地布置井位,降低钻井成本,减少勘探时间,从而减少总体勘探费用。这项技术可以减少钻干井的数量,从而有利于经济和环境。此外,对未来油藏产量预测准确度增加证明了对天然气最终采收率的影响很大。事实上,使用四维地震成像技术可使天然气采收率达到65%,甚至70%;而三维地震成像技术只达到50%。

在过去的十年里,三维和四维地震反射剖面法取得了进步,使油气行业探测出更多的潜在及已探明的页岩气和其他非常规天然气储量。水平钻井和水力压裂技术的进步也离不开这些先进勘探技术的引用,而水平钻井和水力压裂等核心技术的发展又促使了页岩气生产经济可行。尽管如此,地震勘探通常要辅助以磁法和重力勘探数据,以及从探井中获得的实际物理数据。

除了用地震学以外,还可以通过测量地下岩层磁性来获取地质和地球物理数据。这就需要用磁力仪来完成了,即用来测量地球磁场各种变化(因地层多种多样)的仪器。此外,地球物理学家还可以测量并记录地球重力场的细微差异,以便更好地了解地下地质特征。通过测量地层的这些差异,地球物理学家能够评估地下岩层的性质,并确定其潜在的可采页岩气储量。

## ⌑ Key to Exercises

Ⅰ. 1. Exploration is a critical initial process step in the development of oil and gas resources. Before shale gas can be extracted from deep beneath the earth's surface through drilling, stimulation, and extraction processes, precise estimates of the location and quantity of potential gas reserves and the geologic and geophysical properties of the subsurface must be developed to maximize the economic viability of subsequent steps and eventual gas production.

2. Seismology is the study of the movement of seismic waves through the earth's crust and their varied interaction with different types of underground formations.

3. The use of seismic, magnetic, and gravity surveys.

4. Advances in 3 – D and 4 – D reflection seismic profiling have enabled the industry to successfully identify and grow potential and proved reserves of shale and other unconventional gas over the past decade.

5. By using 4 – D seismic technologies, finding costs are reduced as exploration teams can more readily identify natural gas prospects, place wells more effectively, reduce drilling costs, and cut exploration time.

Ⅱ. 1. B   2. D   3. A   4. A   5. A   6. C   7. C   8. A   9. D   10. A

Ⅲ. 1. A   2. B   3. A   4. D   5. C

Ⅳ. 1. 反射地震学是人工产生地震波,然后抓取震波在地下的反射,以便于多维呈现地下地质状况。

2. 检波器将接收到的数据随后传送给地震波记录车存储,留待地球物理学家和石油资源工程师对其作进一步解释。

3. 高等电子学理论的应用使接收仪信号基本上实现数字化了,数据可以本地记录,也可以通过雷达或者光纤转至基站。

4. 在过去的十年里,三维和四维地震反射剖面法取得了进步,使油气行业探测出更多的潜在及已探明的页岩气和其他非常规天然气储量。

5. 四维地震技术在三维地震技术基础上增加了时间作为第四维,这样可以观察到地下地质特征在时间上的变化。

Ⅴ. 1. 勘探是油气资源开发过程中关键性的第一步。在地层深处开发页岩气,首先要精确估算其潜在储量的位置和质量,并确定气藏的地质和地球物理特性,然后再进行钻井、增产并开采等步骤,最终实现页岩气生产的经济可行性和产量最大化。

2. 三位地震勘探与二维地震学的原理相同,只不过是用的地震检波检器更多,将其布置成网格状,从而生成颗粒更小的地下三维视图。从20世纪90年代起,三维地震技术的应用逐步增多,虽然三维勘探成本更高些,但精确定位高产区域的概率大幅增加,足以抵消其成本费用。

## 4.2 钻井

### 导语

钻井通常分为两大类型:探井和开发井。探井主要的目的是为了确定地下岩层是否有油气存在。只有钻到目的层才能确定是否有油气存在。钻井可以垂直(垂直井)或以一定角度(定向井)钻进,定向钻井可以躲避某些特殊地形特征到达目的层。定向钻井和水平钻井常用来进入附近区域地下目的层,减少对气田开发的地面破坏,加大气井和产层的接触面积。

### 课文

**勘探井**

尽管远程和地面勘探技术不断进步,但想更全面地了解地下地质情况,并确定页岩气的存在和可采储量还需要钻勘探井。地质学家通过检测钻屑和泥浆,可以更充分地了解地下地质特征。地层评价或测井是开发井和勘探井中常用的一项技术。

为了充分掌握页岩储层的潜在特点,需要对很多探井做分析,特别是地质盆地较大且目标页岩区域变化多端的地方。由于钻井作业耗时长、成本高,所以只有在其他数据都表明进入页岩气储层概率较大时才钻探井。

**井场建设**

筹备页岩气井场的第一步是修路和井场建设。井场地表用来安放钻机钻井以及完井后生产。修路是为了提供油井建设和维护的通道。钻一口垂直井只需建造一个井场以及相关的道路即可,而一个井场及通道也可同时支持多个水平钻井。

修建通道得先清理好路线,并备好路面以便运送重型设备。地面清理好后即可铺上一层碎石,然后压实,铺设防腐材料,并修建排水沟及涵洞等以利于排水。道路修建的规模取决于现有道路至井场的距离,道路因躲避某些地貌特征所需要选择的路线,以及井场建设和钻井过程中需要运送至井场设备的大小。

井场建设的准备工作主要包括:清理和平整出适当的区域以放置钻探设备、泥浆池和水池等,并铺好地面以方便大型设备的搬运。地面准备工作通常是在清理好的区域铺一层碎石,然后压实,铺设防腐材料,再修建泥浆池和水池。此外,还要建立页岩气集输系统,连接每个单井和输气管道,输送天然气。

不包括公用设施和道路占地在内,井场本身建设一般需要占地 2 ~ 5acre,以便钻井和压裂作业时有足够空间可以搬运设备,安放蓄水池等。天然气生产过程中,如果不再需要实现这些目的,就可以恢复土地,将井场实际占地面积减少到 1 到 3 英亩左右,如图 4.3 所示。

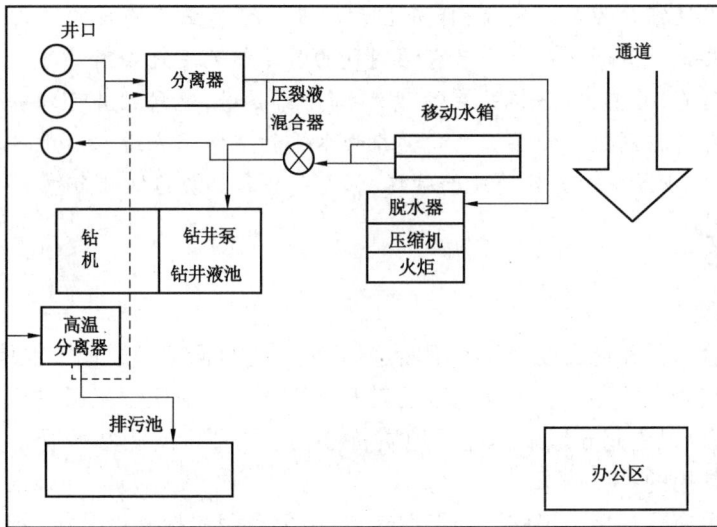

图 4.3　井场示意图

### 钻井

页岩气开采既需要垂直钻井也需要水平钻井。随着技术的进步,直井和水平井的经济性及可行性都有所提高,水平井可以增加井和产层的接触面积,所以更受页岩气开发商的青睐。垂直井和水平井采用的技术和设备类似,但也有一些差异。比如,钻水平井需要使用更大的钻机,还要精心选择钻井液的类型。但水平井钻井最特殊的是使用井下马达,使钻头可以改变方向。

### 垂直钻井

现代天然气开采,无论是水平钻井还是垂直井钻井,通常都采用旋转钻机钻井,如图4.4所示。由空心钻杆连接成的钻柱,将动力系统的扭矩传递到旋转的加压钻头,钻穿地表和地下沉积物。钻头往往用工业钻石镶嵌,能磨碎任何类型的岩石。

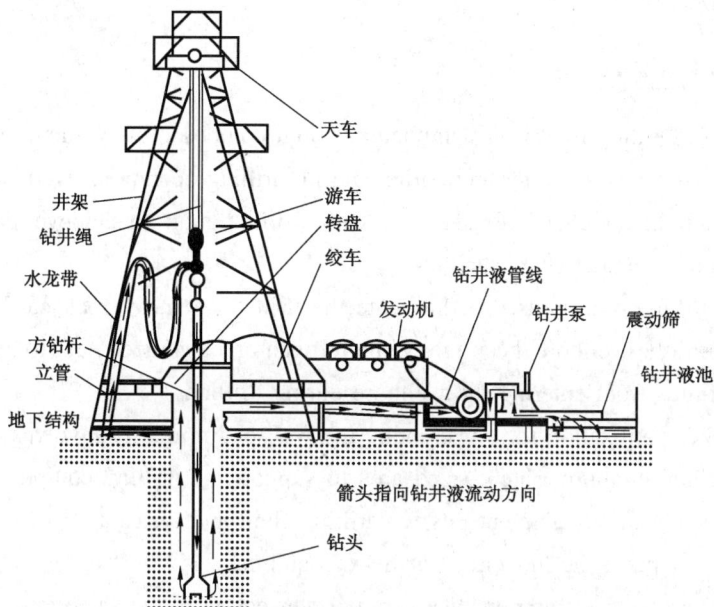

图4.4 旋转钻井示意图

钻井液,即钻井泥浆,通过钻柱泵至钻头起润滑作用,把岩屑等废物携带出井筒,保持井眼开放和稳定,并在钻井过程中防止地层流体侵入井筒。钻井液是由合成油、聚合物基流体、或者水和重晶石制成。在钻井作业中,井场工程师经常把化学添加剂加入钻井液中,使钻井液具备特殊的物理或化学性质。

钻井液携带岩屑,从钻柱和井壁间的环形空间返到地表的振动筛或传送带上,过滤掉岩屑。然后钻井液返回到钻井液池,处理后循环使用。同时,钻头不断钻进,钻柱一段段地续接,直到钻达目的深度。当钻头钻入气藏时,常要调整钻井液成分,使其最大限度地减少对储层的伤害。

**水平钻井**

定向钻井可使井眼轨迹偏离传统的垂直井眼。尤其适合页岩气开采,因为页岩气储层通常呈现为水平方向。水平钻井可使井筒下部或水平部分井筒与页岩气储层平行交叉。

钻水平井可增加井与气藏的接触面,更有利于采出束缚在岩石里面的天然气,从而提高生产效率。另外,水平井还能够开采到垂直井难以开采的资源,例如位于人口稠密地区或处于湖泊等地质结构下面的页岩气藏。页岩气开发中水平层段的井筒长度可在 1000~5000 英尺。此外,水平钻井可实现在一个井场钻多个分支井,使勘探开发公司减少了井场建设成本和地面麻烦。

## Key to Exercises

Ⅰ. 1. Drilling mud is a manufactured fluid composed of synthetic oils, polymer-based fluids, or water and barite. During drilling operations, well site engineers often blend chemical additives with drilling mud to achieve particular physical or chemical characteristics.

2. Drilling mud is used to lubricate the drill bit, remove rock cuttings and waste from the wellbore, keep the wellbore open and stabilized, and to keep formation fluids from entering the wellbore during drilling.

3. Site preparation for constructing the well pad primarily involves clearing and leveling an appropriately sized area to support the drilling equipment footprint and related containment pits for drilling fluid and water and preparing the surface to support the movement of heavy equipment.

4. Horizontal drilling enables parallel intersection of the shale formation by the lower part, or horizontal lateral, of the wellbore, as well as enables increased pay zone contact.

5. Drilling a horizontal well requires the use of larger drilling rigs and the selection of types of drilling fluid used. However, the key difference in horizontal drilling is the use of downhole motors to enable the drill bit to change direction.

Ⅱ. 1. A  2. D  3. A  4. D  5. A  6. B  7. B  8. C  9. B  10. D

Ⅲ. 1. B  2. B  3. B  4. C  5. A

Ⅳ. 1. 道路修建的规模取决于现有道路至井场的距离,道路因躲避某些地貌特征所需要选择的路线,以及井场建设和钻井过程中需要运送至井场设备的大小。

2. 此外,还要建立页岩气集输系统,连接每个单井和输气管道,输送天然气。

3. 垂直井和水平井钻井采用的技术和设备类似,但也有一些差异。

4. 在钻井作业中,井场工程师经常把化学添加剂加入泥浆中,使泥浆具备特殊的物理或化学性质。

5. 钻水平井可增加井与气藏的接触面,更有利于采出束缚在岩石里面的天然气,从而提高生产效率。

Ⅴ.1. 不包括公用设施和道路占地在内,井场本身建设一般需要占地 2~5 英亩,以便钻井和压裂作业时有足够空间可以搬运设备,安放蓄水池等。天然气生产过程中,如果不再需要实现这些目的,就可以恢复土地,将井场实际占地面积减少到 1~3 英亩。

2. 泥浆携带岩屑,从钻柱和井壁间的环形空间返到地表的泥浆振动筛或传送带上,过滤掉岩屑。然后泥浆返回到泥浆池,处理后循环使用。同时,钻头不断钻进,钻柱一段段地续接,直到钻达目的深度。当钻头钻入气藏时,常要调整泥浆成分,使其最大限度地减少对储层的伤害。

## 4.3　水平钻井与测井方法

### 导语

水平钻井先垂直钻井到产层一端,然后在垂直约 90°处开始造斜,再沿产层水平钻进。与传统可转向动力钻具相比,旋转转向系统克服了停钻和启动上的技术困难,可在钻头持续旋转的同时导航钻头,能更精确地控制钻进轨迹,节省钻井时间。测井是在井筒内用探测器测量岩石及岩石内流体的物理和化学特性,而泥浆录井是通过分析带到地面上的地层岩芯样本和钻屑,或通过钻井过程中的其他线索,如钻进速度等,得到很多地下信息。

### 课文

**传统可转向动力钻具**

水平钻井中方向变化是由可转向动力钻具实现的。可转向动力钻具组件安装在钻柱上,是井下组装设备(BHA)的一部分。BHA 由钻头、给钻头加压的钻铤以及保持组合钻具处于井眼中心的平衡器组成。可转向动力钻具组件包括动力部分(可将钻井液水动力转变为旋转运动)、高达 3°的弯曲部分(可实现非垂直钻头方向)、驱动轴以及钻头本身。

使用传统可转向动力钻具进行定向钻井有两种模式:旋转(旋转模式)和定向(也称滑动模式)。在旋转模式下,钻柱旋转穿透岩石。如要改变井眼方向,在电机组件弯头指向新轨迹时,使钻柱停止旋转,然后再开动钻进。

用传统可转向动力钻具进行水平钻井会产生一些问题,使钻井和完井效率低下。旋转停止时,井下岩屑和钻井液的巨大重力会使钻柱卡在井里。旋转模式和滑动模式交替使用会导致通向气层的井眼路线扭曲以及井眼内部粗糙。这样就会延长完井和下套管的时间。

### 旋转转向系统

旋转转向系统(RSS)是在20世纪90年代中期实现商业化生产的,该系统配运用高精度测量仪器,使定向钻井更加经济、有效,是过去十年里页岩气快速产增的关键因素。传统可转向动力钻具在改变井眼方向时,须先停止旋转,待轨迹确定后再启动,而旋转转向系统在转弯时钻头可以旋转不停,这是较之传统可转向动力钻具先进之处。

在RSS系统中,换向系统是工程师和地质学家共同设计的井眼轨迹程序。转向动力钻具安装在钻头后面的钻柱上。井下设备装有传感器,以便连续记录和报告探测数据(如深度、倾角和方向等),这样地面操作者就能追踪到钻头的确切位置。钻井液传输的脉冲使系统导航钻头沿着既定井眼轨迹钻进,或者根据需要修正方向。

旋转转向系统分为两类:"推式钻头"和"摆动钻头"。推式钻头在井下组装设备中装有偏置器和控制器。钻柱旋转时,偏置器会按照指定方向用力推动钻头。控制器由自动力电子器件、传感器及一个控制钻头力度和侧载方向的装置构成,可实现预定轨迹。摆动钻头可使钻头按预定方向倾斜来控制钻头方向。

旋转转向系统可使操作者在钻头持续旋转时控制钻头方向,更好地控制轨迹。与传统可转向动力钻具相比,旋转转向系统克服了停钻和启动上的技术困难,减少了钻井时间,并提高了井筒位置的精确度。这些改进都给运营商带来了经济效益。

在页岩气开发中,钻井效率(以每米总成本计算)是关键的经济因素。转向时连续旋转技术使钻进效率更高,同时该转向模式更加稳定,可更好地控制井眼轨迹,也使井筒内壁更光滑。旋转转向系统克服了许多传统可转向动力钻具中的常见难题,可使井眼更加准确地沿着产层延伸(2500～9000ft),提高了钻井的经济效益。

### 测井

地层评价是测量和分析井眼周围岩石的物理性质,以确定地层边界、碳氢化合物储量和可采能力。这一过程也叫测井,能详细记录井眼周围的地质信息(测井记录),记录油气井开发不同阶段的情况,包括钻井、完井、生产和油井废弃等。测井曲线可以基于对带到地表的地下岩层样品目测所得的地质录井数据,也可基于下入井中测井仪进行物理测量所得的地球物理测井数据。

### 泥浆录井

地质录井数据来源与地表,而不是井下仪器。油气行业最常用的地质录井技术是泥浆录井,通过检测旋转钻井过程中钻井液循环上来的小块岩石来做出录井曲线。页岩气开发中最受关注的泥浆录井参数是页岩气储量。

除了岩性(岩石样本的物理特性)之外,泥浆录井图通常还包括实时钻井参数,如钻进速度、钻井液温度和氯化物等。泥浆录井数据可能还包括钻井液比重、孔隙压力、方向或井斜数据、钻压、转速、泵送率和压力、黏度、钻头参数和套管鞋深度等。

### 电缆测井

在地球物理测井中,将电子传感器下入井筒测量岩石密度、孔隙度、渗透率和电阻率。沿井筒下入成像工具拍摄地层图像。通过声波测井可测量岩性并与之前的地震调查数据比对。通过磁共振测量可显示地层中流体的体积和类型,也可直接测出地层渗透率。核测井仪器可用伽马射线轰击地层岩石,能够准确测出渗透率,而且适用于装有生产套管的井,因此在页岩气开发中特别有用。核磁共振(NMR)测井仪器是一种新技术,用于测量岩石孔隙度,根据孔隙大小分布计算渗透率,并识别孔隙内流体(如水、石油和天然气)。

最常见的两种地球物理数据收集方法是电缆测井和随钻测井(LDW)。电缆测井是在钻井完成后沿井筒下入测量仪器的,可用于评价油气井完整性,评价油气井下套管过程中注水泥情况,并可用于生产井的生产测井。

### 随钻测井

在随钻测井中,测井仪器安装在钻柱上,随钻井进程进行测量。随钻测井是实时进行的,在油藏还未受到钻进侵害之前测量,因此数据更准确。实时测量还能有助于地质导向,LDW 工具识别的地质标记可用于调节钻头的钻进方向,引导井眼钻至特定目标位置。

提供这种定向信息的随钻测井工具也叫随钻测量(MDW)工具。在高级测井系统中,MDW 工具安装在旋转转向系统里,按照预设井眼轨迹自动调节钻进方向。实时测量、测井和分析井下地球物理数据的能力,以及确保井眼位于最佳油气资源位置的能力,是开发低渗透率深层页岩气资源,成功进行水平钻井的关键因素。

## Key to Exercises

Ⅰ. 1. Several issues leading to drilling and well completion inefficiencies are associated with the use of conventional steerable motors for horizontal drilling. When not rotating, the drill string can become stuck under the substantial weight of downhole cuttings and drilling mud. Alternation between rotating and sliding modes results in both a more tortuous path to the gas formation and rougher wellbore interior. These issues increase the time required to complete drilling and casing operations.

2. Nuclear logging instruments can actively bombard rocks with gamma rays to enable accurate estimates of permeability and can be used in wells with production casings.

3. Rotary steerable systems (RSS), commercialized in the mid – 1990s, combined with highly accurate measuring devices, have enabled directional drilling to become both more effective and economical and are key enablers of the rapid growth in shale gas production over the past decade. These systems enable the drill bit to continue rotating while changing angle or steering-an advancement over the stop and start operation of conventional steerable motors.

4. RSS can be categorized into two types: "push-the-bit" and "point-the-bit" systems. Push-the bit systems incorporate bias and control units into the BHA. The bias unit applies force to "push" the bit in a controlled direction while the drill string rotates. The control unit contains self-powered electronics, sensors, and a mechanism to control the degree of force and direction of the bit side loads to achieve the desired trajectory. Point-the-bit systems steer the bit by tilting it in the direction of the desired angle.

5. In addition to lithology (physical characteristics of rock samples), mud logs also typically include real-time drilling parameters such as rate of penetration, temperature of the drilling fluid, and chlorides. Data logged during the process may also include mud weight, pore pressure, directional data or deviation surveys, weight on bit, rotary speed, pump rate and pressure, viscosity, drill bit parameters, casing shoe depth, and formation tops amongst others.

Ⅱ. 1. A   2. B   3. A   4. A   5. A   6. D   7. B   8. A   9. B   10. A

Ⅲ. 1. A   2. A   3. B   4. D   5. C   6. B   7. B   8. A   9. B   10. A

Ⅳ. 1. 旋转模式和滑动模式交替使用会导致通向气层的井眼路线扭曲以及井眼内部粗糙。

2. 与传统可转向马达相比,减少了停钻和启动上的技术困难,减少了钻井时间,并提高了井筒位置的精确度。

3. 如要改变井眼方向,在电机组件弯头指向新轨迹时,使钻柱停止旋转,然后再开动钻进。

4. 用传统可转向马达进行水平钻井会产生一些问题,使钻井和完井效率低下。

5. 通过磁共振测量可显示地层中流体的体积和类型,也可直接测出地层渗透率。

Ⅴ.1. 地质录井数据来源与地表,而不是井下仪器。油气行业最常用的地质录井技术是泥浆录井,通过检测旋转钻井过程中泥浆循环上来的小块岩石来做出录井曲线。页岩气开发中最受关注的泥浆录井参数是页岩气储量。

2. 最常见的两种地球物理数据收集方法是电缆测井和随钻测井(LDW)。电缆测井是在钻井完成后沿井筒下入测量仪器的,可用于评价油气井完整性,评价油气井下套管过程中注水泥情况,并可用于生产井的生产测井。

# 4.4 完井

## 导语

在油气开发中,完井就是完成油气井建设的最后作业过程,使油气井具备油气生产条件。具体任务是按照规定完成井底建设工作,下生产套管和油管,以及井下配套设备,还须按要求射孔和增产。页岩气开发一般使用套管射孔完井法。然而,完井技术的进步促使了水平井裸眼多段压裂技术出现,裸眼完井技术也已广泛应用于页岩气开发。而增产处理具体方法取决于地层类型,水力压裂法可用于页岩地层开发,提高页岩气流动速度。

## 课文

### 下套管

套管用于防止井内气体泄漏至周围岩层,以及周围岩层中的水涌入井里。套管益处很多,最主要的是确保储层中天然气的采收和出售,同时防止周围淡水层污染。套管还可以使上部地层井眼保持稳定以免坍塌,封闭地表防止高压地层发生井喷,同时使井筒内壁平滑,可以安装生产油管和设备。

下套管是在井眼中安装一层层的同心大直径钢管,并注水泥。每安装一层套管,就要注水泥以确保地层与套管之间以及两层套管之间封隔完好。每层套管都有其特殊用途。气井套管通常包括:

- 导管(最外层的套管)可阻止近地表松散土壤坍塌,并在钻井过程中循环钻井液。导管直径通常为 16 ~ 20in,长度约 50ft,并在钻井作业开始前就要下入。

- 表层套管用以阻止钻井液和采出气污染近地表的淡水层;其长度取决于淡水层深度。
- 中间套管用于特殊区段,保护油气井免于各种地下危险情况,如压力异常地区和盐水层等。
- 生产套管(最里层套管)为油管导管,油管最后下入到生产套管内,将采出气输送到地面。

为防止含水层和井眼间发生窜槽,在井深 1000 ~ 4000ft 下入表层钢套管并注水泥。一旦钻过潜水面,就要起钻,下表层套管;然后把套管和井壁之间的环形空间用水泥注满。水泥凝固后,从第一段注好水泥的套管底部继续钻进到下一个深度。随后还要下中间套管,每次下套管都重复上述程序,但是套管直径一次比一次小,直至钻至页岩气产层,下入生产套管。

套管注入水泥后要进行注水泥评估。用电子探测器测量水泥厚度,确保其能达到设计效果。在美国,套管和水泥规格以及下套管和注水泥的流程都是由各州和联邦明文规定。此外,测试生产套管是为确保套管能承受住压力,容纳水力压裂过程中使用的压裂液。一旦这些测试成功完成,接下来就是油井生产准备工作:完井和增产。

**完井和增产**

完井和增产作业可使天然气从地层沿井筒流到地面。完井基本上是按设计规格进行井底作业以实现产气量最大化,安装油管和相关井下设备,射孔并增产(总称"低部完井"),同时安装井口设备。增产是对含烃地层进行处理,以提高气体由地层到井筒的流速。增产处理具体方法取决于地层类型,水力压裂法可用于页岩地层开发,提高页岩气流动速度。

**完井方法**

油气开发完井技术有几好种。完井技术的选择主要取决于采气方法的技术和经济可行性,也要考虑气藏的位置和特点。一般而言,以下两种技术会有一种用于页岩气完井:

- 套管射孔完井,先下套管和注水泥至井底,再用射孔弹对产层井壁射孔,然后进行水力压裂。
- 裸眼完井,裸眼完井不用在产层下套管注水泥,井底裸露,气体通过生产尾管流入井筒。

页岩气开发一般使用套管射孔完井法。然而,完井技术的进步促使了水平井裸眼多段压裂技术出现,裸眼完井技术也已广泛用于美国页岩气开发。裸眼完井能够降低下套管、注水泥和射孔等成本。而且,与常规射孔完井相比,裸眼完井能够提高页岩气藏天然气的初次采收率,但不太容易控制裂缝分布和路径。

在进行压裂作业之前,要先安装井口设备,即为水力压裂专门设计的"压裂树"。此外,还要在地表安装管汇和气水分离器等返排设备,并对系统进行压力测试。压裂结束后,用高压井口装置替换压裂树,监测并调节流出井口的天然气及其他流体。

### 水力压裂

页岩由于孔隙度和渗透率都低,无法释放出其中的气体。因此,页岩完井采用水力压裂法来提高页岩气采收率。水力压裂方法也称为压裂,即把产层井筒分隔成段,然后沿井筒高压泵入大量特制压裂液(高达 8000psi),压裂液从井筒出去进入页岩层。流体压力撑开裂缝并使裂缝继续延伸。

裂缝使暴露于井眼中的地层表面积增加,也建立起连接产层和井筒的通道,提高了储层流体的生产率,有时能提高数百个百分点。一定大小颗粒的沙子类支撑剂随压裂液进入裂缝,使裂缝张开,使气体不断流动。为了使储层接触面积最大化,井筒水平段都要以最佳方式增产。对有天然裂缝的页岩进行压裂处理,是为了把天然裂缝互相连通,建立通向井筒的路径。

在某些页岩地层进行套管完井,成本高,不经济。每段压裂都需要使用多孔射孔枪及挠性管起下作业,因此,需要反复安装和拆卸增产设备。近年来裸眼多段完井在技术上更具可行性,省去了注水泥、射孔和分隔成本,并且通过减少压裂作业时间降低了人员及设备成本。

### Key to Exercises

Ⅰ. 1. There are several benefits. Chief among them is ensuring that the gas accessed from the reservoir can be recovered for sale and protecting surrounding formations such as fresh water zones from contamination. Casing also provides stability to the upper formations of the wellbore to prevent it from caving, seals the surface from high pressure zones to prevent blowout, and creates a smooth internal bore for installing production tubing and equipment.

2. The typical gas well casing consists of conductor casing, surface casing, intermediate casing and production casing.

3. The two techniques used in shale gas well completions are:

• Cased-hole, perforated completion, where production casing and cement extend to the toe(bottom tip) of the well, the walls of which are perforated in the production zone using a shaped explosive charge prior to hydraulic fracturing.

• Open-hole completion, in which no casing is cemented into place within the reservoir, the end of the piping is left open and gas is collected through the production liner.

4. Open-hole completions enable the elimination of costs associated with casing, cementing, and perforating. Further, they have been found to enable higher initial gas production rates from shale reservoirs than conventional perforated completions.

5. Completion and stimulation activities enable the flow of natural gas out of the formation and up to the surface.

II. 1. C  2. B  3. A  4. A  5. A  6. B  7. C  8. D  9. A  10. A

III. 1. A  2. A  3. B  4. A  5. C

IV. 1. 用电子探测器测量水泥厚度,确保其能达到设计效果。

2. 完井和增产作业可使天然气从地层沿井筒流到地面。

3. 页岩由于孔隙度和渗透率都低,无法释放出其中的气体。

4. 压裂结束后,用高压井口装置替换压裂树,监测并调节流出井口的天然气及其他流体。

5. 增产处理具体方法取决于地层类型,水力压裂法可用于页岩地层开发,提高页岩气流动速度。

V. 1. 套管用于防止井内气体泄漏至周围岩层,以及周围岩层中的水涌入井里。套管益处很多,最主要的是确保储层中天然气的采收和出售,同时防止周围淡水层污染。套管还可以使上部地层井眼保持稳定以免坍塌,封闭地表防止高压地层发生井喷,同时使井筒内壁平滑,可以安装生产油管和设备。

2. 裂缝使暴露于井眼中的地层表面积增加,也建立起连接产层和井筒的通道,提高了储层流体的生产率,有时能提高数百个百分点。一定大小颗粒的沙子类支撑剂随压裂液进入裂缝,使裂缝张开,使气体不断流动。为了使储层接触面积最大化,井筒水平段都要以最佳方式增产。对有天然裂缝的页岩进行压裂处理,是为了把天然裂缝互相连通,建立通向井筒的路径。

# 第5章 生产对环境的影响

## 5.1 压裂用水

### 导语

目前对页岩气藏进行水力压裂处理的方法基本上是将几百万加仑混有支撑剂和增稠剂的水基压裂液分阶段监控泵入页岩目标地层。用于增产的压裂液主要由水构成,但也包括各种化学添加剂,可能会带来污染。压裂处理过程中添加剂的用量一般取决于压裂井的具体情况。在水源紧张地区,钻井和压裂用水可能会造成很多严重的环境问题,如地下水位下降、生物多样性影响以及当地生态系统的破坏。

### 课文

水用量及污染风险是目前所有非常规天然气开发中的关键问题,已引发社会广泛关注。对页岩气或致密气藏的开发来说,虽然钻井要用水,但是用水最多的是水力压裂的过程:每口井大概需要几千到两万立方米(相当于 $100 \times 10^4 \sim 500 \times 10^4\, \mathrm{gal}$)的水。水的有效利用在水力压裂中非常重要。从 2010 年中期开始,美国西得克萨斯伊格福特区页岩开发区每口井完井平均用水量从 $18500\mathrm{m}^3$ 减少到 $13600\mathrm{m}^3$,主要是因为增加了压裂液返排水的循环利用,这是页岩气开发向前迈进的重要一步。2011 年,德州铁路委员会又颁发了 2800 多口伊格尔福特油井开采许可证。若以单位能量生产的耗水量来计算,页岩气或致密气开发用水量比常规气藏开发多,但和常规油藏开发用量相近。

压裂用水可以是地表水源或现场水井(为支持生产钻好的,从浅水层或深水层抽水),或者从更远的地区用卡车运水。将水从水源运输到工地堪称工程规模浩大。如果一口井水力压裂需要 $15000\mathrm{m}^3$ 的水,卡车运载量一般为 $30\mathrm{m}^3$,那么就需要 500 辆卡车运输。这么大的运输量会导致当地道路堵塞,并加速道路和桥梁的磨损,如果管理不当的话,会增加交通事故。

在水源匮乏地区,钻井和压裂用水可能会造成很多严重的环境问题,如地下水位下降、生物多样性影响以及当地生态系统的破坏;同时还有可能减少当地居民用水量和其他生产活动如农业用水量。

在缺水地区,有限的压裂用水严重限制了致密气和页岩气的开发。例如在我国,新疆维吾尔自治区塔里木盆地拥有最大的页岩气藏,但那里水资源极度匮乏,使页岩气开发面临困难。其他一些比较有前景的页岩气开发区虽不及塔里木地区资源储量丰富,水源也没那么紧张,但水资源竞争也异常激烈。中国页岩气的开发之所以集中于四川盆地,部分原因是该盆地水源比较充足。

非常规气井对淡水的需求取决于水力压裂,页岩气开发所采用的主要压裂液"滑溜水"通常成本最低,在某些页岩气储集层中也能起到增产的作用,但其耗水量最高。因此,可减少压裂用水量的方法备受人们关注。使用传统高粘性压裂液(添加聚合物或表面活性剂),可以降低总注水量(需水量),但这需要添加复杂的化学制剂。

借助于表面活性剂(如洗涤液中使用的),用氮气或二氧化碳使水汽炮,就制成了泡沫压裂液。这种压裂液比较有吸引力,因为90%为气体,具有很好的支撑剂携带性能。虽然使用烃基压裂液,如丙烷或稠化烃,确实能减少耗水量,但这些化合物具有可燃性,在井场使用很危险。返排压裂液的比例与使用的压裂液种类(以及页岩性质)有关,因此压裂液的最佳选择需要考虑诸多因素:水是否容易获得、水是否循环利用、页岩气藏的特性,化学制剂的使用是否要减少,以及其他经济因素。

**什么是压裂液?**

人们关注的环境问题主要是水力压裂过程中使用的压裂液以及压裂液渗入地下水中所造成的水污染。一般压裂液99%的成分是由水和沙砾或陶瓷粒(即"支撑剂")构成,为使压裂液具备一定性能,还需使用化学添加剂。地层类型不同压裂液所需性能也不同,添加剂(并非所有压裂都需要)通常有以下四种作用:

• 向井内泵入压裂液时,使支撑剂悬浮于其中,并确保支撑剂最终进入撑开的裂缝中。否则,支撑剂会因颗粒较重,将在重力作用下在压裂液中不均匀分散,无法起到应有的作用。瓜尔豆和纤维素(类似于食品与化妆品中所用的)等的胶凝聚合物在压裂液中使用比例约为1%。硼酸盐和金属盐之类的交联剂也常微量用于增加压裂液的胶凝性。矿物水中此类物质天然含量往往较少,但浓度高时,这些物质具有毒性。

• 随着时间的推移改变压裂液特性。将支撑剂带入地下裂缝的性能在压裂过程中的其他阶段并不需要,因此有些添加剂对压裂液的性能具有时效性。比如,压裂完成后,压裂液黏度降低可使烃气更易于沿裂缝流入井中。通常在后期使用工业用氧化物或者酶等低浓度的螯合剂(如用于水壶去垢的)破坏胶凝聚合物。

● 减小摩擦力,从而减小将压裂液注入井中所需要的动力。通常使用的一种减阻剂叫做聚丙烯酰胺(应用广泛,比如在婴儿纸尿裤中起到吸水作用)。

● 减少水中细菌对压裂液性能的影响,阻止细菌在产层中繁殖,产生硫化氢。通常需要使用消毒剂(抗微生物剂),类似于医院或生活中常用的消毒用品。

## Key to Exercises

Ⅰ. 1. The amount of water required forunconventional gas developments is higher.

2. The development of China's shale gas industry has to date focused on the Sichuan basin, in part because water is much more abundant in this region.

3. Hydraulic fracturing dominates the freshwater requirements for unconventional gas wells.

4. In foamed fluids, water is foamed with nitrogen or $CO_2$, with the help of surfactants. Foamed fluids can be attractive, as 90% of the fluid can be gas and this fluid has very good proppant-carrying properties.

5. Water itself, together with sand or ceramic beads (the "proppant"), makes up over 99% of a typical fracturing fluid, but a mixture of chemical additives is also used to give the fluid the properties that are needed for fracturing.

Ⅱ. 1. B    2. C    3. C    4. A    5. D    6. D    7. A    8. A    9. B    10. D

Ⅲ. 1. B    2. C    3. D    4. A    5. B

Ⅳ. 1. 水用量及污染风险是目前所有非常规天然气开发中的关键问题,已引发社会广泛关注。

2. 矿物水中此类物质天然含量往往较少,但浓度高时,这些物质具有毒性。

3. 在缺水地区,有限的压裂用水严重限制了致密气藏和页岩气的开发。

4. 返排压裂液的比例与使用的压裂液种类(以及页岩性质)有关,因此压裂液的最佳选择需要考虑诸多因素:水是否容易获得、水是否循环利用、页岩气藏的特性、化学制剂的使用是否要减少,以及其他经济因素。

5. 人们关注的环境问题主要是水力压裂过程中使用的压裂液,以及压裂液渗入地下水中所造成的水污染。

Ⅴ. 1. 在水源匮乏地区,钻井和压裂用水可能会造成很多严重的环境问题,如地下水位下降、生物多样性影响以及当地生态系统的破坏;同时还有可能减少当地居民用水量和其他生产活动如农业用水量。

2. 压裂用水可以是地表水源或现场水井,或者从更远的地区用卡车运水。将水从水源运输到工地堪称工程规模浩大。如果一口井水力压裂需要15000m³ 的水,卡车运载量一般为30m³,那么就需要 500 辆卡车运输。这么大的运输量会导致当地道路堵塞,并加速道路和桥梁的磨损,如果管理不当的话,会增加交通事故。

## 5.2 压裂液添加剂

### 📇 导语

因为要满足每个开发区的特殊需求,每种压裂液的配制及添加剂用量都不尽相同。压裂液一般分为三种:水基、油基和泡沫压裂液。因为成本低、性能高、悬浮力强,对环境影响不大又易于使用,水基压裂液已广泛应用。油基压裂液易燃,比较危险,主要用于水敏性地层。泡沫压裂液一般用于渗透率和压力较低的页岩地层,可以减少地层伤害,避免裂缝堵塞。但其支撑剂携带能力差,设备成本高,和水基以及油基压裂液相比很不经济。

### 📇 课文

直到最近,压裂液的化学成分一直是商业机密,未公之于众。由于公众坚持认为有权力知晓注入到自己土地中的物质究竟是什么,再把压裂液成分作为商业机密已然不合时宜。自 2010 年以来,主动披露压裂液成分在美国大部分地区已成为惯例。而且油气行业还设法在不使用可能有害的化学制剂的情况下,获得理想的开发成果。在美国,用滑溜水做压裂液越来越普遍。"滑溜水"由水、支撑剂、普通减阻聚合物以及抗微生物剂制成,但泵送速度要求高,且只能携带细粒支撑剂。此外人们还注意尽量减少意外地面泄露,大多数专家认为这是污染地下水的最大威胁。

抗微生物剂、表面活性剂、防垢剂等化学添加剂,也可作为压裂液的成分。抗微生物剂,如溴、甲醇或萘,可以抑制细菌生物的繁殖。细菌生物会导致裂缝堵塞以及井眼缩颈。异丙醇等表面活性剂可用于改变压裂液的表面张力,而乙二醇则用于防止管内结垢。其他添加剂,如胶凝剂、交联剂和破胶剂可用来控制流体粘度,并使流体能根据需要携带或释放支撑剂。

压裂液中可能有的添加剂类型及其作用如表 5.1 所示。用于压裂处理的添加剂数量、类型和浓度要根据具体井况和地层条件而定,其配方通常为压裂液生产商所专有。但总体上讲,大多数滑溜水压裂液中添加剂的浓度都在 0.5% ~2% ,水在溶液中的比例占 98% ~99.5% 。

**表5.1　水力压裂液添加剂的类型与作用**

| 类型 | 作用 | 常用化学品举例 |
|---|---|---|
| 酸剂 | 注入压裂液前清除掉射孔段的残留水泥与泥浆 | 盐酸(浓度3%~28%) |
| 抗微生物剂 | 抑制微生物生长,微生物可能会污染甲烷气,影响流体流动 | 戊二醛;溴硝醇 |
| 破胶剂 | 降低流体粘度,释放支撑剂至裂缝,使压裂液易于返排 | 过二硫酸盐 |
| 黏土稳定剂 | 防止地层黏土膨胀运移,导致渗透率下降 | 盐(如四甲基氯化铵和氯化钾) |
| 缓蚀剂 | 抑制含酸压裂液对钢制油管、套管、储罐及其他设备所造成的锈蚀 | 甲醇 |
| 交联剂 | 交联各类制剂,以增加压裂液黏度 | 氢氧化钾 |
| 减阻剂 | 使摩阻最小化,使压裂液注入速度和压力最大化 | 丙烯酸丙烯酰胺共聚物钠 |
| 胶凝剂 | 增加压裂液黏度,使更多支撑剂进入裂缝 | 瓜尔豆胶 |
| 铁量控制 | 防止金属氧化物沉淀,避免堵塞地层 | 柠檬酸、巯基乙酸 |
| 支撑剂 | 将裂缝撑开,使天然气更易于从地层流入井眼 | 石英砂、烧结铝矾土、陶瓷粒 |
| 阻垢剂 | 防止碳酸盐和硫酸盐沉淀,避免地层堵塞 | 氯化铵、乙二醇、聚丙烯酸酯 |
| 表面活性剂 | 降低压裂液表面张力,利于其返排 | 甲醇、异丙醇 |

### 泡沫压裂

在某些页岩气藏压裂中,可用泡沫压裂液替代滑溜水压裂液。该技术使用惰性气体,一般是氮气或二氧化碳,与发泡剂和水混合后制成泡沫,携带支撑剂至页岩裂缝并沉淀。较之高压滑溜水压裂,泡沫压裂主要具有两大优势:一是降低了地层损害,避免造成裂缝堵塞;二是大大减少了压裂用水量。

减少地层注水量可以降低黏土膨胀、运移及裂缝和井眼堵塞的可能性,减少返排后残留在地层中的水量,同时提高产量。另外,水力压裂作业中注入及排出井眼的工业用水量也会大大减少。如此一来可以节约成本,因为用于水运输、储存和处理的设备以及相关费用都减少了。

泡沫压裂技术一般用于渗透率和压力较低的页岩地层,对于这样的地层尤其要注意防止诱发性损害,以及黏土运移导致的堵塞。这项技术兴起于数十年前,现已成功用于阿巴拉契亚盆地。然而,在巴奈特页岩气藏早期开发中,发现部分区域并不适合使用泡沫压裂技术,这些区域的脆性石英岩用高压滑溜水压裂液效果最好,也最经济。随着生产人员不断研究提高压裂液稳定性和粘度的方法,泡沫压裂技术也在不断进步,携带支撑剂的性能进一步提高,压裂液损失逐步降低。在条件适宜的气藏,包括阿肯色州的费耶特维尔页岩气藏和加拿大西部的沉积盆地气藏,用泡沫压裂液技术取代滑溜水压裂液技术受到越来越多的关注。

另一种适合用于低压、水敏性页岩气藏的压裂技术是氮气或二氧化碳气体压裂。该方法使用无水压裂液,如100%的氮气(或二氧化碳),携带特殊设计的轻质支撑剂。如果不使用水悬浮液,可以降低高压水力压裂导致的毛细作用和地层损害。使用轻质支撑剂还可以减少地层开裂和细小颗粒的产生,从而提高气体流向井筒的能力。

上面讨论的各种技术突破比以往页岩气藏的发现、开发和生产都更有效。美国页岩气主产区使用这些技术生产效率最高,使美国页岩气产量在过去的十年里大幅增长。实际上,页岩气使美国这个世界上最大的天然气消费国在2009年转变为最大的天然气生产国。

专家们认为页岩气在世界各大洲的储量极大,加上页岩气在美国开发的巨大成功,掀起了全球开发页岩气的热潮。由于常规天然气开发较成熟,且其开发相对容易,能源安全与独立曾一度遭到忽视,影响了世界各国开发非常规油气资源的兴趣。现在几大油田服务公司正在使美国的页岩气开发技术快速推向全球。但这些技术并不像常规油气开采那样成熟,仍需继续完善,应对各种独特页岩地层的特殊挑战。

## 🔳 Key to Exercises

Ⅰ. 1. Reducing accidental surface spills is regarded as a more significant risk of contamination to groundwater by most experts.

2. Overall the concentration of additives in most slickwater fracturing fluids is relatively consistent 0. 5 percent to 2 percent with water making up 98 percent to 99. 5 percent of the solution.

3. The two key advantages of foam fracturing over high-pressure slickwater fracturing are reduction of formation damage that could lead to blocked fractures and substantial reduction of the volume of water that is needed for the fracturing.

4. Nitrogen or carbon dioxide fracturing is suitable for underpressurized, water-sensitive shales.

5. The use of light-weight proppants can reduce formation spalling as well as generation of fine particles, and increase conductivity of the gas to the wellbore.

Ⅱ. 1. D  2. B  3. B  4. B  5. D  6. A  7. B  8. D  9. B  10. C

Ⅲ. 1. A   2. D   3. B   4. B   5. B

Ⅳ. 1. 这样可以节约成本,因为用于水运输、储存和处理的设备以及相关费用都减少了。

2. 抗微生物剂、表面活性剂、防垢剂等化学添加剂,也可作为压裂液的成分。

3. 用于压裂处理的添加剂数量、类型和浓度要根据具体井况和地层条件而定,其配方通常为压裂液生产商所专有。

4. 泡沫压裂技术一般用于渗透率和压力较低的页岩地层,对于这样的地层尤其要注意防止诱发性损害,以及黏土运移导致的堵塞。

5. 美国页岩气主产区使用这些技术生产效率最高。

Ⅴ. 1. 直到最近,压裂液的化学成分一直是商业机密,未公之于众。由于公众坚持认为有权力知晓注入到自己土地中的物质究竟是什么,再把压裂液成分作为商业机密已然不合时宜。自2010年以来,主动披露压裂液成分在美国大部分地区已成为惯例。而且油气行业还设法在不使用可能有害的化学制剂的情况下,获得理想的开发成果。

2. 在某些页岩气藏压裂中,可用泡沫压裂液替代滑溜水压裂液。该技术使用惰性气体,一般是氮气或二氧化碳,与发泡剂和水混合后制成泡沫,携带支撑剂至页岩裂缝并沉淀。较之高压滑溜水压裂,泡沫压裂主要具有两大优势:一是降低了地层损害,避免造成裂缝堵塞;二是大大减少了压裂用水量。

# 5.3   油田废水的处理与处置

## 导语

常规油气和非常规油气开发都会导致潜在的水源污染。运输泄漏、套管断裂、压裂岩石漏失、产层烃类或化学物质通过中间岩石向浅水层运移、井场地面排放、压裂液返排、煤层气生产排水以及废水处理都会导致水污染。最大的污染危险来自废水处理。废水处理包括但不限于下面这几种方法:注入地下排放,处理后排放至地面水体或者在水力压裂作业中循环使用。

## 课文

**水力压裂生产中的废水**
对于非常规天然气生产来说,尤其是在水力压裂用水量极大的情况下,废水的处理与处置至关重要。将压裂液注入井眼后,在随后的几天到几周里,部分压裂液(通常几乎全是水)会以反排液的形式返回至地面。返排液总量取决于地质情况。页岩地层通常会返出注入量的20% 50%,其余的则与页岩中黏土结合,遗留在地层中。反排液中含有一些水力压裂过程中所使用的化学物质,另外还有金属、矿物质和储层岩石中滤出的烃,含盐量通常很高。有些气藏中滤出的矿物可能略具放射性,因此需要在地表采取特殊的防护措施。钻井废水之类的返排流体需要就地安全储存,且最好全部储存在牢固的、防风雨的储藏设备中。若不妥善处置,这些流体将对当地环境造成潜在威胁。

压裂废水分离后,有多种方法可以处理。最佳方法是将其回收留待日后利用,这在技术上完全可行,不过出于经济方面的考虑,还不能随时都有水循环用于水力压裂。其次是在当地工业污水处理厂将废水净化,使其达到一定标准后排入当地河流,或用于农业生产。再者,如果地层条件合适,可以直接将废水注入更深的地层。

**煤层气生产中的采出水**

煤层气开采过程中基本不需要额外供水,但是,妥善处理排水过程中井里采出的水十分重要。采出水可以重新注回地下隔离层,排入现有排水系统,或排放到浅池塘蒸发,如果净化妥当,还可以用于灌溉和其他生产活动。选择何种方法处理采出水取决于诸多因素,其中水质尤为重要。根据不同的煤层沉积环境和水文条件,采出水可能盐、钠含量很高(含高浓度的钠、钙、镁),另外还可能含有微量有机化合物,因此通常需要处理后才能用于农业灌溉等。使用盐水灌溉会抑制植物发芽与生长,而水中钠含量过高会改变土壤的物理性质,导致排水不良及地表结壳,最终影响农作物的产量。

废水处理成本由处理程度和处理量决定。实际上,每口井因产气所需排出的总水量有很大不同。例如,在蒙大拿和怀俄明州的粉河盆地,估计每口煤层气井的日排出水量平均为 $65m^3$($1.7 \times 10^4$gal)。据估计,2008 年美国从煤层中排水总共 $1.8 \times 10^8 m^3$($470 \times 10^8$gal),相当于旧金山一年的直接用水量。

原则上,采出水可以净化到任意所需标准。这样做虽然成本高,但只要运输成本合理,用于生产还是能带来经济效益的。采出水处理方法的选择以及附近水的市场价格是决定煤层气开发经济效益的关键因素。

现今许多煤层气生产地区,或者生产前景良好的地区,常常处于干旱及半干旱环境中,如果有外界淡水供应的话,将会非常受益。由于净化采出水用于灌溉或将其重新注回更深地层代价高昂,因此北美目前最普遍的做法仍然是使其蒸发或排入废水系统(有时会处理后排放),而在澳大利亚,处理后的水越来越多地循环使用了。2008 年,美国总采出水中的 45%,即大约 $8500 \times 10^4 m^3 (220 \times 10^8 gal)$,几乎没有进行任何处理,就排入了地表水中。

对煤层气采出水所造成的实际环境影响进行评估的经验是有限的。一份美国国家研究委员会最近的研究成果显示,采出水的最终处理和使用对土壤、生态系统、地表水及地下水的水质和水量所造成的影响既有正面的也有负面的。尽管该研究没有发现大量负面证据,但我们必须承认这一事实,即煤层气行业刚刚起步,对给当地所造成的影响的详细调查还太少。

**水源污染风险**

水源潜在污染的危险已经引起了人们的极大关注,无论是地表水源(如河流或者浅淡水层),还是地层更深处由于各类非常规气藏开采所产生的地下水源。造成水源污染的因素主要有以下四个:

① 流体或固体意外泄漏于地表,包括钻井液、压裂液、水或采出水、烃类物质以及固体废物。

② 套管注水泥过程中,因为水泥密封不好造成压裂液漏失以及地层深处的盐水或烃向浅水层渗漏。

③ 产层的烃类物质或化学添加剂通过中间的岩石孔隙向浅水层渗漏。

④ 将未处理完善的废水排入地下。

这些危害因素并不只存在于非常规油气开采中,常规油气的开发,无论是否进行水力压裂,同样存在这些危险。然而,非常规资源的开发规模不可避免地增加了事故发生的概率。

公众已将关注点聚焦于第三个污染源,也就是产层的烃类或化学物质可能通过中间岩石向浅水层移动。然而实际上,这可能是诸多危害中影响最小的一个,至少在页岩气或致密气藏的生产中是这样。在某些情况下,将关注点放在这上面就可能会将监管者的视线或时间从其他更紧迫的问题上转移走。

### Key to Exercises

I. 1. Flow-back water contains some of the chemicals used in the hydraulic fracturing process, together with metals, minerals and hydrocarbons leached from the reservoir rock.

2. The optimal solution is to recycle it for future use.

3. Excessively sodic water can change the physical properties of the soil, leading to poor drainage and crusting and adversely affecting crop yields.

4. It is because of the high cost of purifying the water for irrigation or reinjection into a deeper layer.

5. The potential cost of water disposal depends on both the extent to which treatment is required and the volume of water produced.

Ⅱ. 1. B　2. C　3. D　4. B　5. D　6. C　7. B　8. B　9. B　10. A

Ⅲ. 1. C　2. B　3. C　4. C　5. C

Ⅳ. 1. 对于非常规天然气生产来说,尤其是在水力压裂用水量极大的情况下,废水的处理与处置至关重要。

2. 煤层气开采过程中基本不需要额外供水,但是,妥善处理排水过程中井里采出的水十分重要。

3. 采出水可以重新注回地下隔离层,排入现有排水系统,或排放到浅池塘蒸发,如果净化妥当,还可以用于灌溉和其他生产活动。

4. 这样做虽然成本高,但只要运输成本合理,用于生产还是能带来经济效益的。

5. 一份美国国家研究委员会最近的研究成果显示,采出水的最终处理和使用对土壤、生态系统、地表水及地下水的水质和水量所造成的影响既有正面的也有负面的。

Ⅴ. 1. 选择何种方法处理采出水取决于诸多因素,其中水质尤为重要。根据不同的煤层沉积环境和水文条件,采出水可能盐、钠含量很高,另外还可能含有微量有机化合物,因此通常需要处理后才能用于农业灌溉等。使用盐水灌溉会抑制植物发芽与生长,而水中钠含量过高会改变土壤的物理性质,导致排水不良及地表结壳,最终影响农作物的产量。

2. 公众已将关注点聚焦于第三个污染源,也就是产层的烃类或化学物质可能通过中间岩石向浅水层移动。然而实际上,这可能是诸多危害中影响最小的一个,至少在页岩气或致密气藏的生产中是这样。在某些情况下,将关注点放在这上面就可能会将监管者的视线或时间从其他更紧迫的问题上转移走。

# 5.4　温室气体排放

🔲 导语

地球大气层中的主要几种温室气体是水蒸气、二氧化碳、甲烷、一氧化氮和臭氧。自工业革命起,化石燃料的燃烧和原始森林的大量砍伐导致了大气中二氧化碳浓度增加40%。甲烷是天然气系统泄漏,以及牲畜饲养等人类活动致使的第二大温室气体。甲烷在大气中的存在时间比二氧化碳短得多,但其对气候的影响每 100 年比二氧化碳高 20 倍。

## 课文

页岩气和致密气生产比常规气生产排放的温室气体多。主要原因有以下两方面:

① 每生产一立方米的天然气,就需要钻更多的井以及进行更多的水力压裂。这些作业都需要消耗能量,而能量通常来源于柴油发动机,因而导致单位有用能量的生产释放更多的二氧化碳。

② 完井过程中需要排放和点燃更多的气体。在压裂液返排期间,每口井总采出量有所提高(因为水力压裂作业多,压裂液返排需要时间更长,而岩石渗透率又低,导致每口井的总采收量通常少一些)。

为了观察压裂液返排过程中气体排放和燃烧这两种做法会产生什么不同的结果,我们根据环保局相关数据,发布了对这两种影响所做的评估。气体燃烧排放比常规气生产高出 3.5%,但如果是气体直接排放,这一数据将升至 12%。根据“黄金准则”,做到零气体排放、最大限度降低天然气燃烧量,采收并销售返排过程中产生的天然气,可以降低这一数据。

煤层气开发中温室气体排放问题同样受到关注,完井时的水力压裂以及排水到煤层气生产过程中,都会有大量甲烷排入大气中。重要的是在钻井、压裂及生产作业中,要进行精心管理与操作,将气体排放量降至最低。这需要用特殊设备把天然气从采出水或压裂液中分离出来,然后将气体注入集气系统(或临时储存设备)。如果出于技术、后勤或者经济上的原因难以做到的话,最好将气体烧掉,而不是将其排出,因为这样做安全且产生的温室效应要小得多。

对天然气在生产、运输和使用过程中的温室气体排放问题,以及非常规气生产中的其他排放问题,一直都有些争议。有人认为对天然气(尤其是非常规天然气)作为一次能源使用而产生的废气排放问题估计严重不足。还有人甚至认为,非常规天然气整个生产过程中的废气排放要高于煤炭生产。甲烷排放的主要问题不仅在于天然气生产,还在于其运输及使用。

甲烷产生的温室效应要强于二氧化碳,但其在大气中存在的时间要相对短些——二氧化碳半衰期为 150 年,而甲烷半衰期只有 15 年。因此,比较甲烷和二氧化碳对全球产生的温室效应有几种不同的方法。一种方法是对比二氧化碳,估计甲烷 100 多年的全球变暖潜值(GWP)。据政府间气候变化专门委员会(IPCC)的第四份报告,这100年的GWP为25,在上一份报告GWP为21的基础上进行了修正。从消除目前甲烷排放源所带来的长期相关利益来看,这一数值很说明问题。

据IPCC估测,约二十多年后GWP值便会达到72。可以说这一数字更接近将来二三十年甲烷的大量排放估值,是决定世界能否达到控制地表平均温度增长不超过2℃这一目标的关键。还有些科学家认为,甲烷与空气悬浮颗粒的相互作用也会增大甲烷的GWP,可能会使GWP更高。

估值如此之高不仅对天然气产业链的甲烷排放有所启示,对其他方面的甲烷排放也一样,如牲畜甲烷排放、垃圾场甲烷排放、稻田及其他农业生产甲烷排放,另外还有大自然。

天然气产业链(包括常规与非常规)的甲烷排放值主要源于以下四个方面:

① 出于安全或者经济方面的考虑,故意排放天然气。完井排气即属此类,设备保养也会有排气发生。

② 天然气漏失。这种情况通常发生于管道、阀门或密封泄漏,既有偶发因素(如管道腐蚀),也有人为设计因素(如旋转密封,敞口储存罐)。

③ 封闭设备破裂等意外情况。包括管道、压力罐及井口隔离装置等。

④ 不完全燃烧。天然气火炬燃烧完全情况根据风力等其他各种情况不尽相同,通常不会超过98%。(如同点燃煤气炉一样:通常几秒钟后火苗才能稳定)。

鉴于天然气的这些特性,我们很难量化其排放。大部分估算都是美国环保局和天然气研究协会依据产业链中各个部分(气井、设备及管道等)的排放系数,在20世纪90年代做出的研究成果。但这决不表明这些研究可以说明世界其他区域的排放情况,或者将来甲烷排放的可能状况。估计全球天然气产业链的甲烷排放量会占天然气产量的1%~8%。因为没有更好的方法,在2011年美国环保局对未来甲烷排放的综合预测中,排放系数没有变化,对2010—2030年油气行业的预测是甲烷排放量增加26%。

关于甲烷排放量及其对环境影响程度的不同估测,会对天然气是"清洁"化石燃料这一看法产生极大的影响。因为天然气每单位能量燃烧所排放的二氧化碳少,人们普遍认可天然气对煤炭的优势;但是很显然悲观一点的估测又会使天然气成为比煤更差的温室气体排放源。重要的是,我们还需要不断地进行科学研究来找到最接近的GWP值,并且需要投入双倍精力更加系统地测量甲烷的排放量。

## Key to Exercises

Ⅰ. 1. More wells and more hydraulic fracturing are needed per cubic metre of unconventional gas produced. These operations use energy, typically coming from diesel motors, leading to higher $CO_2$ emissions per unit of useful energy produced. In addition, more venting or flaring is needed during well completion. The flow-back phase after hydraulic fracturing also represents a larger percentage of the total recovery per well.

2. The gas must be separated from the produced water with specialized equipment before it is injected into a gasgathering system.

3. There are different possible ways to compare the effect of methane and $CO_2$ on global warming. One way is to evaluate the Global Warming Potential (GWP) of methane, compared to $CO_2$, averaged over 100 years.

4. The main issue revolves around methane emissions not only during production, but also during transportation and use of natural gas. Methane emissions can also result from livestock, landfills, rice paddies and other agricultural sources, as well as from natural sources.

5. The advantage of gas over coal is widely accepted when it is assumed that $CO_2$ emissions per unit of energy from gas combustion is lower; but more pessimistic assumptions can make gas a worse greenhouse-gas emitter than coal.

Ⅱ. 1. C  2. B  3. D  4. A  5. C  6. D  7. D  8. C  9. C  10. C

Ⅲ. 1. C  2. D  3. C  4. D  5. A

Ⅳ. 1. 为了观察压裂液返排过程中气体排放和燃烧这两种做法会产生什么不同的结果,我们根据环保局相关数据,发布了对这两种影响所做的评估。

2. 根据"黄金准则",零气体排放、最大限度降低天然气燃烧量,采收并销售返排过程中产生的天然气,可以降低这一数据。

3. 还有人甚至认为,非常规气整个生产过程中的废气排放要高于煤炭生产。

4. 从消除目前甲烷排放源所带来的长期相关利益来看,这一数值很说明问题。

5. 但这决不表明这些研究可以说明世界其他区域的排放情况,或者将来甲烷排放的可能状况。

Ⅴ. 1. 重要的是在钻井、压裂及生产作业中,要进行精心管理与操作,将气体排放量降至最低。这需要用特殊设备把天然气从采出水或压裂液中分离出来,然后将气体注入集气系统(或临时储存设备)。如果出于技术、后勤或者经济上的原因难以做到的话,最好将气体烧掉,而不是将其排出,因为这样做安全,且产生的温室效应要小得多。

2. 对天然气在生产、运输和使用过程中的温室气体排放问题,以及非常规气生产中的其他排放问题,一直都有些争议。有人认为对天然气(尤其是非常规天然气)作为一次能源使用而产生的废气排放问题估计严重不足。还有人甚至认为,非常规气整个生产过程中的废气排放要高于煤炭生产。甲烷排放的主要问题不仅在于天然气生产,还在于其运输及使用。

# 第6章 非常规油气市场

## 6.1 生产成本和天然气价格

### 导语

历史上天然气国际贸易非常有限,因为天然气是当地生产、当地消费。天然气价格机制包括政府指导定价、与具竞争力的燃料挂钩定价,主要是石油产品或者是市场现货定价。因为地理位置和市场情况的差异,世界各地区天然气生产成本有所不同,因而定价也不一致。基于天然气各种生产成本,可以估算出其盈亏平衡成本(等同于其商业收入的生产成本)。采用盈亏平衡方法有利于更好地进行成本效益分析和定价。

### 课文

非常规天然气开发的成本包括:资本成本、运营成本、税收和特许权使用费。

资本成本,通常称为勘探和开发成本,主要用于油气井建设。同样情况下,页岩气井比常规天然气井成本要高,这是由于多段水力压裂作业费用额外高;同样,致密气井成本较高也是这个原因。与常规天然气井相比,煤层气井目前成本费用较低,因为煤层气生产一直是在市场较成熟的地区进行,且产层较浅。运营成本,也称采收成本,是和生产活动直接相关的非固定成本:运营成本可能会根据当地条件的不同而有所变化。将天然气运往市场的成本取决于距离的远近,在这方面常规和非常规天然气是相同的。

最后是税收和特许权使用费,在不同的司法管辖区这些费用也有差别;除了利润税,税收和特许权使用费通常还包括支付给政府的固定或生产性税费,以及/或支付给资源所有者(可能是也可能不是政府)的特许权使用费。由于地理或市场条件不同,资本和运营成本较高的国家或地区,通常会建立更具吸引力的财政制度吸引投资,有时甚至可能提供补贴:中国对煤层气和页岩气生产就采取了补贴政策。

基于上述成本费用,就可以预测"盈亏平衡成本"或"供应成本",即足以回报新项目投资的市场价值。采油成本、运输成本、税收和特许权使用费通常是以每单位气体生产所需美元直接来表示的。资本成本的高低更是取决于每口气井的采收量。气井产量高低差别很大:据报道,美国最好的页岩气井预计最终采收率(EUR)是$(1.5 \sim 3) \times 10^8 m^3 (50 \sim 100) \times 10^8 ft^3$;但是很多页岩气井的最终采收率都要低于该数字$10 \sim 100$倍。每口页岩气井的平均预计最终采收率都不相同,当然这个采收率也取决于开发商针对特定页岩的行业经验:随着时间的推移,页岩气产业不断优化技术,能从每口井中开采出更多的天然气。到目前为止,美国以外地区基本上没有经验,但是钻长水平井应有助于提高每口井的最终采收率(在美国许多管辖区,水平井长度受限于面积尺寸的规定)。

因此,天然气的盈亏平衡成本在不同区域或在一国之内都有很大不同。比如在美国,干气井的盈亏平衡成本大概在5美元/MBtu(百万英热单位)到7美元/MBtu的范围内;含有液体的天然气盈亏平衡成本要低一些,可低达3美元/MBtu,因为液体虽然使成本增加了一点,但却增加了很大的价值(油井伴生气盈亏平衡成本更低)。

由于陆上常规天然气资源已接近枯竭,未来常规天然气产量将来自成本更高的海上,因此,在美国常规和非常规天然气盈亏平衡成本范围基本上是一致的。

欧洲的生产成本要高出50%,盈亏平衡成本会在5美元/MBtu到10美元/MBtu之间。常规和非常规天然气成本差别不大,因为常规资源枯竭,新项目正迁至成本更高的挪威北极地区。中国的成本结构和美国类似,但是中国的页岩埋藏更深,地质结构更复杂;同样,中国的煤层气藏也很偏远,所以中国的盈亏平衡成本范围可能处于美国和欧洲之间,在4美元/MBtu到8美元/MBtu之间(不过有生产补贴,这个数值还可以减小)。对中国的这一估值既适用于常规天然气也适用于非常规天然气,因为容易开采的常规天然气临近枯竭,生产正转向海上或更偏远的地区。在那些仍然拥有大量相对容易开采的常规天然气国家,如中东地区国家,盈亏平衡成本还不到2美元/MBtu,但其非常规天然气盈亏平衡成本会相对高一些(类似于美国的非常规天然气)。

北美地区非常规天然气产业发展最快,因此对该地区天然气市场及价格的影响也最大。相对于其他能源形式(如石油),天然气价格创下历史新低。更令人惊讶的是,北美市场与其他主要天然气消费市场相对隔离,页岩气开发已产生了深远的国际影响。这是因为北美已基本实现天然气自足,而在2000—2010年的十年间,人们认为北美地区未来将成为庞大的液化天然气净进口地区,因此进行了许多液化天然气投资。

在此期间,仅美国就建了超过$1000 \times 10^8 m^3$的进口基础设施,同时在卡塔尔等主要产气国进行液化气供应配套设施投资。然而 2011 年,北美液化气净进口量还不足 $200 \times 10^8 m^3$,而全球液化气市场交易量超过了 $8500 \times 10^8 m^3$:$80 \times 10^8 m^3$ 输往美国,$90 \times 10^8 m^3$ 输往墨西哥和加拿大。因此,大量液化天然气可以供给其他市场了,包括亚洲和欧洲。

目前,在世界许多市场,包括北美、英国及澳大利亚部分地区,天然气实行自由定价,这种机制称为气与气之间的竞争。然而,亚太地区许多跨国天然气交易都是长期合同,与油价或炼制产品价格挂钩。欧洲大陆天然气价格主要与油价挂钩,不过最近几年两个系统有所交叉(两者间有很多不同),在与油价挂钩的同时,天然气定价还要进行气与气之间的竞争,过程复杂。能源安全、地缘政治以及向环保型燃料的转变(应对气候变化的关键),都将取决于将来天然气的定价方式。

## Key to Exercises

Ⅰ. 1. The costs of developing and producing unconventional gas are made up of several elements: capital costs, operational costs, transportation costs, and taxes and royalties.

2. Because production has been at shallow depths in regions with well-developed markets.

3. Becausesome countries or regions have higher capital and operating costs, due to their geography or market conditions, a more attractive fiscal regime is created in order to attract investment.

4. Because the North American region was expectated to be a substantial net LNG importer.

5. An approach in which gas prices are set freely is known as gas-to-gas competition.

Ⅱ. 1. C  2. C  3. D  4. A  5. B  6. C  7. C  8. B  9. C  10. C

Ⅲ. 1. B  2. C  3. C  4. C  5. A

Ⅳ. 1. 运营成本,也称采收成本,是和生产活动直接相关的非固定成本:运营成本可能会根据当地条件的不同而有所变化。

2. 基于上述成本费用,就可以预测"盈亏平衡成本"或"供应成本",即足以回报新项目投资的市场价值。

3. 常规和非常规天然气成本差别不大,因为常规资源枯竭,新项目正迁至成本更高的挪威北极地区。

4. 对中国的这一估值既适用于常规天然气也适用于非常规天然气,因为容易开采的常规天然气临近枯竭,生产正转向海上或更偏远的地区。

5. 每口页岩气井的平均预计最终采收率都不相同,当然这个采收率也取决于开发商针对特定页岩的行业经验:随着时间的推移,页岩气产业不断优化技术,能从每口井中开采出更多的天然气。

Ⅴ.1. 资本成本,通常称为勘探和开发成本,主要用于油气井建设。同样情况下,页岩气井比常规天然气井成本要高,这是由于多段水力压裂作业费用额外高;同样,致密气井成本较高也是这个原因。与常规天然气井相比,煤层气井目前成本费用较低,因为煤层气生产一直是在市场较成熟的地区进行,且产层较浅。

2. 在此期间,仅美国就建了超过 $1000 \times 10^8 m^3$ 的进口基础设施,同时在卡塔尔等主要产气国进行液化气供应配套设施投资。然而 2011 年,北美液化气净进口还不足 $200 \times 10^8 m^3$,而全球液化气市场交易量超过了 $8500 \times 10^8 m^3$:$80 \times 10^8 m^3$ 输往美国,$90 \times 10^8 m^3$ 输往墨西哥和加拿大。因此,大量液化天然气可以供给其他市场了,包括亚洲和欧洲。

# 6.2 矿产权许可与租赁

## 导语

获取采矿权是页岩气开发过程中的关键步骤。如果没有采矿权,就不能合法开采并销售天然气。世界各地矿权所有制的不同造成了各地油气行业迥异的发展现状,对各国经济产生了重大影响,同时也使美国页岩气勘探与开发工艺和技术得以早早领先。美国和世界其他地区矿权所有制的不同是其页岩气生产领先世界的一个重要因素。

## 课文

矿场租赁协议是矿主和石油公司之间的协议,石油公司可以在协议规定的范围内获得矿藏勘探、开发以及开采的权利。矿产权可以归政府公有,也可以归公司或个人私有。实际上,国家大部分矿产权都掌握在私人手中,而不是政府手中。另外,在矿产所有人同意的情况下,矿产权是可以和地表权分开的,这意味着矿产权可以直接出售或出租。

**政府矿产权**

各国获得政府油气开发矿产权的具体程序都不一样。一般情况下,政府把其所拥有的土地划分为不连续的区块,然后附带各种不同条款和条件,给油气公司颁发许可,允许其勘探并最终开发地下油气资源。通常许可条款包括油气公司进行某些勘探活动需要承担的义务,有时是具体的投资额度。作为交换,油气公司获得数年的勘探权以发现商业油气藏。如果勘探成功了,公司再向政府申请开发许可权。申请通常要有详细的开发方案,包括环境影响说明,以及油气井、工厂和基础设施部署计划。政府会就气田利益进行洽谈,将勘探权转为多年(通常为 15～20 年)开采权。要获得国有矿产开发权很有可能需要数年时间才能完成,其部分原因是世界各地存在着不同程度的官僚作风及其他情况。

### 私有矿产权

私人矿权租赁一般是在签约时(为给承租人保留一定时期内的矿产使用权),即支付矿区租金。如果在租赁的矿区发现了具有商业开采价值的油气储藏,油气公司可能会进一步开发生产。特许权使用费一般包括在矿场租赁协议里,协议确保矿主获得井口气份额。如果租约期满还未开始生产,那么所有的矿产权和财产权都要归还给矿主,除非续约。

如果确认气井最终钻好并且产气,则一次性支付租金(一般以英亩为单位),按井口气产量比例估算特许权使用费。有两个关键因素会影响租赁费及特许权使用费。第一个因素涉及承租人对页岩气生产的信心,以及对矿产租赁的渴求程度,在页岩气开发早期,承租人给出的这笔费用比较有限,因为他们不确定某一区域是否具有商业开采价值,随着开发区块的开采价值日益明朗,天然气开发公司对获得采矿权的渴求程度也越来越强,因此租赁费及特许权使用费也都随之增长。另一个影响特许权使用费的因素是某些特殊地区的租赁监管规章。

### 地产分离

无论是政府矿权还是私有矿权,矿权拥有者不一定都是其地表权拥有者,这就出现了"地产分离"现象,这在美国之外的大多数国家很普遍。在美国,政府和私有矿权的地产分离情况因地区而异。例如,联邦政府拥有西部各州大部分矿产权,在东部和中西部各州,只有 4% 的土地是和联邦政府地产分离的,但这正是美国页岩气资源最丰富的地区。然而,私人矿主之间的地产分离情况也会存在,因为矿权和地表权可以单独交易。

地产分离情况可能引发冲突,因为天然气开发可能要用到矿产之上的地表土地。在这种情况下,矿区承租者有权合理使用地表土地,虽然他们没有地表产权。

### 矿区使用权塑造了美国的页岩气工业

美国页岩气产业有别于世界其他地区的一个因素就是美国和世界其他地区的矿权所有制不同。在美国任何实体都可以随时接洽矿主商定合约,而不是只有在政府颁发了油气勘探许可证之后才可以行动。此外,页岩气生产的利润激发了许多矿主的主动性,使租赁过程大大缩短。在美国,获得矿区租赁权通常只需要几个月,而不像在其他国家那样需要几年。这同时促进了页岩气经济的发展,降低了天然气行业的门槛,为中小型天然气生产商创造了机会。

由于特许权使用费在这些交易中所占份额最大,能让矿主相信自己最有可能快速投产的公司就有优势最早获得矿权。在页岩气开发方面更积极、更高效的公司,比如其产品组合中资源多样性相对有限的独立勘探开发公司,比拥有数十亿美元资产,但不一定会在短期内优先开发页岩气的大公司更加吸引矿主与其达成交易。这是另外一个因素使中小型独立天然气公司在美国页岩气开发中迅速崛起,领先大公司。

### 钻井许可

一旦获得页岩气开发权,开发商便寻求得到钻井许可。在美国,这些许可证由钻井所在的各州发行。各州监管机构有权按照联邦和各州的法律批准并规范页岩气开发活动。气井的设计、位置、间距、操作和遗弃都要由各州监管机构批准,审批过程中需要提供有关方面的详细信息。这些机构还负责颁发以下几个方面的许可:水资源管理、废物处置、废气排放、地下注入、野生动物影响、地表干扰以及工人的健康与安全。最后,其他各级政府,如市、县或者地区的水资源管理部门,除获得各州监管机构授予的许可外,可能还需要其他许可来处理解决注水区井位布置和居民投诉等问题。

### ▣ Key to Exercises

Ⅰ.1. Mineral leases are contractual agreements between the owner of a mineral estate and oil and gas companies to provide the companies the right to explore for, develop, and extract from mineral deposits in the area described in the lease.

2. In the cases of both government and privately owned mineral rights, the owner of the mineral rights on a specific property may be different than the owner of the surface rights, or land that lies above the minerals, which creates a "split estate."

3. Conflict can arise in split estate situations as gas development can require use of the surface land above a mineral estate, which may not belong to the same owner.

4. A deal with a company which appears more efficient and motivated to

develop shale resources, such an independent E&P company with relatively limited resource diversity in its portfolio, can be more attractive to a mineral estate owner than one with a multi-billion dollar supermajor that may or may not prioritize development in the near term.

5. In the United States, the permits to authorize drilling of wells are issued by the individual states in which the well is to be drilled.

Ⅱ. 1. A　2. C　3. B　4. C　5. D　6. B　7. B　8. B　9. B　10. D

Ⅲ. 1. A　2. D　3. B　4. B　5. B

Ⅳ. 1. 矿场租赁协议是矿主和石油公司之间的协议,石油公司可以在协议规定的范围内获得矿藏勘探、开发以及开采的权利。

2. 另外,在矿产所有人同意的情况下,矿产权是可以和地表权分开的,这意味着矿产权可以直接出售或出租。

3. 各国获得政府油气开发矿产权的具体程序都不一样

4. 地产分离情况可能引发冲突,因为天然气开发可能要用到矿产之上的地表土地。

5. 气井的设计、位置、间距、操作和遗弃都要由各州监管机构批准,审批过程中需要提供有关方面的详细信息。

Ⅴ. 1. 在美国,页岩气生产的利润激发了许多矿主的主动性,使租赁过程大大缩短。获得矿区租赁权通常只需要几个月,而不像在其他国家那样需要几年。这同时促进了页岩气经济的发展,降低了天然气行业的门槛,为中小型天然气生产商创造了机会。

2. 无论是政府矿权还是私有矿权,矿权拥有者不一定都是其地表权拥有者,这就出现了"地产分离"现象,这在美国之外的大多数国家很普遍。在美国,政府和私有矿权的地产分离情况因地区而异。例如,联邦政府拥有西部各州大部分矿产权,在东部和中西部各州,只有4%的土地是和联邦政府地产分离的,但这正是美国页岩气资源最丰富的地区。然而,私人矿主之间的地产分离情况也会存在,因为矿权和地表权可以单独交易。

# 6.3　市场需求驱动因素

## 导语

因为天然气比煤效率高,碳排放量少,所以将成为越来越重要的发电燃料。最近由于页岩气供应量增加,气价下降,进一步推动了气代煤的燃料转换。预计世界油价在未来的几十年里继续攀升,因此在可能的情况下都会以价格相对便宜的天然气取代石油。尽管基于目前技术,只有25%的页岩气可开采,但按照目前的天然气消费水平,仅页岩气就能满足全球至少40年的天然气需求。虽然常规油气资源在全球分布不均,但广泛分布的页岩资源会使能源进口国提高国内产能,从而提高整体能源安全度。

## 课文

### 全球能源消费增长

全球能源消费预计会以25%的速度增长,从2005年的 $473 \times 10^{15}$ Btu(英热单位)增长到2020的 $590 \times 10^{15}$ Btu以上。全球经济萧条缓和了消费者对商品、服务及相关制造业的需求,同时2009年的全球能源消耗减少了2.2%。自此,非经合组织(经济合作与发展组织)的发展中国家引领了全球经济复苏。还有迹象表明,美国经济衰退已经结束,但因欧洲和日本经济复苏滞后,EIA国际能源展望预测:到2012年世界大部分国家将从经济衰退中恢复,并重新回到长期增长模式。

能源基础设施发达的经合组织国家在历史上一直占据着世界能源消费的最大份额。然而,非经合组织经济体的能源消费增长现已超过经合组织多年。预计在非经合组织国家中,中国和印度能源需求最高,增速最快。中东和中南美洲的非经合组织国家也有望大幅增长。到2020年,非经合组织经济体将占世界能源消费总量约57%。

### 全球天然气需求增长

天然气作为全球发电燃料将越来越重要,因为天然气比煤效率高,碳排放少。在过去的十年中,美国一直用天然气电厂取代燃煤电厂,且态势不减。最近煤价上升,气价下降,主要原因是页岩气供应量增加,进一步推动了气代煤的燃料转换。从2008—2009年,美国燃煤发电量下降了11.6%,致使煤电在总发电量中占比为44.5%,为1978年以来的最低水平。相比之下,2009年天然气发电上升了4.3%,尽管总发电量下降了4.1%,天然气发电份额增加至23.3%,创下1970年以来的新高。美国天然气消费总量中发电份额也迅速增长,从1996年的17%增长到2009年的30%多。页岩气产量上升,增加了天然气供应。因此,世界许多国家都将以气代煤。

除了用来发电外,天然气也普遍用于居民住宅、工业和商业供暖及制冷,另外还可用作石化产品原料以及运输替代燃料。世界油价在未来的几十年里可能还要继续攀升,因此在可能的情况下都会以价格相对便宜的天然气取代石油。工业区新建的石化厂会越来越多地依靠天然气作为原料。此外,压缩天然气替代燃料车的开发和使用可能将使天然气需求大幅增加。

2010—2020 年,全球天然气需求量将继续以每年 2% 的速度增长,因此源于天然气的能量也将进一步增长 22%。非经合组织经济体的消费量以每年 2.8% 以上的速度增长,约是增速为 1.1% 的经合组织国家的三倍。

**经济合作与发展组织常规天然气供应量下降**

虽然天然气需求和消费量持续增长,但常规天然气生产和供应量却在下降,特别是在欧洲和北美的发达经合组织国家中。EIA 预计,从 2008—2020 年,经合组织国家常规天然气产量将下降 11%。而非经合组织发展中国家常规天然气产量将大幅增长,但大部分将用于国内消费,而非出口。另外,即使有剩余产量,将天然气从产量过剩的国家运往天然气短缺的国家也往往难以实现,而且不经济,也容易受到地缘政治因素的制约。因此,虽然北美地区美国、加拿大及墨西哥之间的天然气交易活跃,但北美和中东地区没有管道天然气交易。同样,俄罗斯大量天然气通过管道运往欧洲,但日本却不能通过管道进口天然气。

**页岩气技术进步**

虽然发现世界各地有大量页岩气资源已有数十年了,但非常规资源埋藏深且渗透率低,因此页岩气商业性开发的技术和经济可行性都很有限。美国在过去的十年中成功开发并应用了页岩气勘探、钻井、增产、开采和环境治理等工艺和技术,证明了页岩气生产的可行性。好多项新技术已经应用于页岩气开发中,其中先进的勘探、钻井和增产技术最有影响力。三维、四维地震勘探技术,以及定向钻井和多段水力压裂等技术的应用使美国在过去的十年中页岩气产量大幅增加。

**全球页岩气资源丰富**

全球页岩气资源总量估计超过 $18775 \times 10^{12} \, \mathrm{ft}^3$,如图 6.1 所示。这一数字表明了世界各地页岩地层中估计的天然气总量,目前已经发现和尚未发现的页岩资源均包括在内。尽管基于目前技术,只有 25% 的页岩气是可开采的,但按照目前的天然气消费水平,仅页岩气就能满足全球至少 40 年的天然气需求。

常规油气资源在全球分布不均,大部分资源都集中在一小部分国家里,这种不平衡导致了不断的地缘政治纷争。页岩资源的广泛分布使一些能源进口国提高了国内产能,减少了进口依赖,提高了整体能源安全度。现在将天然气液化后船运比通过管道远距离运输更经济。页岩气大量供应,加上液化天然气运输设施越来越多地使用,在未来几十年很可能会重塑全球天然气市场和能源政治格局。

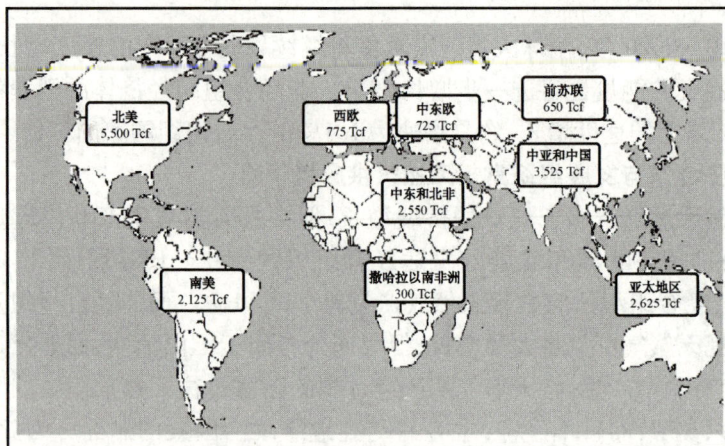

图6.1 全球页岩气资源分布

## Key to Exercises

Ⅰ. 1. China and India are projected to lead energy demand in the non-OECD nations and have the highest growth rates.

2. The pipeline natural gas trade between North America and the Middle East is non-existent. While significant Russian natural gas production is piped into Europe, Japan is unable to import natural gas through pipelines.

3. In addition to its use as a fuel source in electricity generation, natural gas is commonly used as a fuel for residential, industrial and commercial heating and cooling, as a feedstock for various petrochemical products, and as an alternative fuel for the transportation sector.

4. The technological and economic feasibility of recovering commercial quantities of these deep and low-permeability unconventional resources has limited the development of shale gas recovery.

5. The unevenly distributed oil and gas resources around the world will bring about ongoing geo-political strugle.

Ⅱ. 1. C   2. D   3. C   4. A   5. D   6. C   7. A   8. A   9. D   10. B

Ⅲ. 1. B   2. B   3. C   4. D   5. B

Ⅳ. 1. 最近煤价上升,气价下降,主要原因是页岩气供应量增加,进一步推动了气代煤的燃料转换。

2. 世界油价在未来的几十年里可能还要继续攀升,因此在可能的情况下都会以价格相对便宜的天然气取代石油。

3. 非经合组织发展中国家常规天然气产量将大幅增长,但大部分将用于国内消费,而非出口。

4. 虽然发现世界各地有大量页岩气资源已有数十年了,但非常规资源埋藏深且渗透率低,因此页岩气商业性开发的技术和经济可行性都很有限。

5. 常规油气资源在全球分布不均,大部分资源都集中在一小部分国家里,这种不平衡导致了不断的地缘政治纷争。

Ⅴ. 1. 好多项新技术已经应用于页岩气开发中,其中先进的勘探、钻井和增产技术最有影响力。三维、四维地震勘探技术,以及定向钻井和多段水力压裂等技术的应用使美国在过去的十年中页岩气产量大幅增加。

2. 常规油气资源在全球分布不均,大部分资源都集中在一小部分国家里,这种不平衡导致了不断的地缘政治纷争。页岩资源的广泛分布使一些能源进口国提高了国内产能,减少了进口依赖,提高了整体能源安全度。页岩气大量供应,加上液化天然气运输设施越来越多地使用,在未来几十年很可能会重塑全球天然气市场和能源政治格局。

# 6.4　市场需求抑制因素

## 导语

虽然页岩气开发在美国如火如荼,但是页岩气资源勘探和开采仍是一个较新的领域,在美国之外的这些资源的勘探与开发仍处于早期阶段。还有许多诸如开发成本高等挑战需要克服,而且人们在页岩资源可开采规模、生产的经济性、开采过程对环境的影响以及相关的监管政策等方面都有争议。所有这些因素都将抑制页岩气的市场需求。此外,土地和矿产资源开发方面的不利政策也可能会阻碍页岩气行业的发展。

## 课文

**低价天然气有碍页岩气发展**

历史上页岩气钻完井成本一直较高,每囗井的费用需要两三百万美元。为了赚取合理利润,生产商要以更高的价格销售其产品。但要注意的是随着行业生产者不断积累经验,新技术不断提高生产率和降低成本,盈亏平衡价格有望进一步降低,页岩气生产可以获取利润。

如果天然气价格持续低迷,页岩气投资期望值下降,投资额度就可能会受到影响,最终将影响页岩气产量。虽然世界各地区页岩气开发成本和天然气价格都有所不同,但是供求经济学的基本原理是类似的,有可能会影响全球页岩气的投资和增长,这是一个重大风险因素。

**环境问题和监管对策**

对埋藏深且渗透率低的页岩气资源进行持续开采将给环境带来巨大挑战。许多对土地、水和空气等环境的影响都根源于油气勘探和生产中常用的工艺,还有一些影响是页岩气资源开发过程中使用的特殊技术导致的。环境问题对页岩气产业的未来发展构成了风险因素,如果这些问题导致监管力度越来越大的话,更是如此。

在美国,有几个主要联邦法案制约着油气开发对环境等各个方面的影响。这些法规包括:

• 《清洁空气法案》(CAA)限制引擎、天然气处理设备的废气排放,以及其他与钻井和天然气生产相关的废气排放。

• 《清洁水法案》(CWA)规范施工场所地表水排放,以及生产现场的雨水径流。然而,2005年的《能源政策法》将油气井场建设及钻探活动排除在联邦雨水许可条例之外,这种许可由各州自行掌控。

• 《安全饮用水法案》(SDWA)规范油气生产地下液体注入。然而,2005年的《能源政策法》将水力压裂液排除在安全饮水法案联邦条例之外,除非压裂液含有柴油。

• 《国家环境政策法案》(NEPA)要求彻底分析在联邦土地上进行勘探和生产所造成的环境影响。

随着页岩气产量的增加,页岩气开发逐渐转向人口稠密地区,因此环保积极分子在当地有关社团成员的帮助下,提高了页岩气开发技术的风险意识。很多环保激进主义分子支持加强监管力度,认为当地有关部门对页岩气开发的潜在影响监管不够。争论焦点主要是页岩气开发对水的影响,但空气质量和地表破坏也受到了关注。然而,页岩气行业的支持者认为,目前的监管制度很到位,监管力度再加大将对页岩气经济带来负面影响。

**不断发展的监管对策**

由于人们对环境的日益担心,各州已对页岩气生产采取了监管措施,推进了监管行动。具体行动包括提出监管法规,约束废气排放量增多的新设备或改良设备,以及待环境影响研究结束后再启动新的页岩气钻探项目。天然气公司必须遵守这些规定才能获得钻探许可证。例如:

• 2010年7月29日,得克萨斯州环境质量委员会提出了新的规则,确保巴尼特(Barnett)页岩气钻探排放量不会对空气质量产生负面影响。

• 2010年8月16日,怀俄明州州长签发了包括压裂液在内的各种钻井作业新法规,要求生产商提供压裂过程中使用的化学添加剂的详细列表。

● 2010 年 10 月,密歇根自然资源和环境部对需要进行水力压裂的气井提出了特殊要求。

● 自 2010 年 11 月起,纽约州议会几乎不再颁发水力压裂法开发天然气井的许可。

不过天然气行业反对再有联邦监管法规出台,认为其没有必要,因为各州都有自己的监管法规。业内人士还担心遵守联邦法规所支付的成本会给页岩气产业经济造成负面影响。美国独立石油协会的研究表明,联邦法规可能会使每一口新开发的天然气井增加约 10 万美元的成本。环境问题是页岩气产业未来发展的风险因素,如果这些问题致使监管越来越严格的话,尤其如此。

**市场准入障碍**

尽管页岩气地层广泛分布于世界各地,但并不是所有的页岩资源都是一样的。资源区块的地质结构各有不同,甚至同一区块内也经常存在地质条件差异,这都会影响页岩气开发和生产的技术可行性和经济可行性。

像地层深度、厚度、总有机物含量、孔隙度和渗透率等都是决定开发是否会成功的关键因子,但各区域这些参数的差别都相当大。这些差异决定了目前技术的可行性,钻井和压裂的成本、气井产量、产量下降速度以及最终可采天然气的质量和数量。由于北美之外的大多数页岩资源的详细勘探和评估还没有进行,很可能有一部分在某些国家早期进行勘探及评估的资源,现在看来在经济或者技术上是不可开采的。

此外,大部分的技术开发和实践至今都是在美国进行的,因此目前许多希望开发页岩气资源的国家没有具备页岩开发特殊技能和经验的人员。而且美国以外地区的页岩资源开发还受一些其他本土因素的限制,例如现场天然气处理和储存的限制,运输基础设施的限制,以及钻井和压裂专用设备的限制等。

土地和矿产资源开发方面的不利政策也可能会阻碍页岩气行业的发展。获得政府矿产权的具体程序因国而异。通常,政府将土地划分为多个区块,然后给油气公司颁发许可(附带各种不同条款和条件),允许其勘探并最终开发地下油气资源。某些国家给自己的国有开发公司保留或者优先给与其矿产开发许可权,这在中国就是惯例。其他一些国家可能只允许进入指定种类的天然气开发项目,目前的矿产开发许可还没有涉及到页岩项目,像印度就是这种情况。

## Key to Exercises

Ⅰ. 1. If natural gas prices remain low, reduced expectations of shale investments may limit capital expenditures and, ultimately, the rate of growth in shale gas production.

2. Many of these environmental impacts to land, water, and air are inherent to the processes commonly used in oil and gas exploration and production, and yet others result from the specific technologies deployed in the shale gas resource development process.

3. Many of these activists support increased regulatory action, maintaining that localities do not adequately regulate the potential effects of shale gas development. However, shale industry proponents maintain that the current regime of regulatory oversight is adequate and additional requirements would negatively affect the economics of shale gas production.

4. The deciding factor is the geological differences from one resource to another among which critical success factors such as depth and thickness of formation, total organic content, and porosity and permeability can vary widely.

5. The barriers to the growth of shale industry in some countries include the technological difficulties caused by geological conditions, a skillful and experienced workforce for the shale industry, some indigenous factors such as related limitations and devices and lastly the adverse polices regarding land and mineral development.

Ⅱ. 1. B   2. A   3. D   4. B   5. A   6. C   7. B   8. A   9. B   10. A

Ⅲ. 1. B   2. D   3. C   4. B   5. C

Ⅳ. 1. 对埋藏深且渗透率低的页岩气资源进行持续开采将给环境带来巨大挑战。

2. 由于人们对环境的日益担心,页岩气生产各州已采取了监管措施,推进了监管行动。

3. 土地和矿产资源开发方面的不利政策也可能会阻碍页岩气行业的发展。

4. 由于北美之外的大多数页岩资源的详细勘探和评估还没有进行,很可能有一部分在某些国家早期进行勘探及评估的资源,现在看来在经济或技术上是不可开采的。

5. 然后政府给油气公司颁发许可(附带各种条款和条件),允许其勘探并最终开发地下油气资源。

Ⅴ. 1. 历史上页岩气钻完井成本一直较高。为了赚取合理利润,生产商要以更高的价格销售其产品。但要注意的是随着行业生产者不断积累经验,新技术不断提高生产率和降低成本,盈亏平衡价格有望进一步降低,页岩气生产可以获取利润。

2. 很多环保激进主义分子支持加强监管力度,认为当地有关部门对页岩气开发的潜在影响监管不够。争论焦点主要是页岩气开发对水的影响,但空气质量和地表破坏也受到了关注。然而,页岩气行业的支持者认为,目前的监管制度很到位,监管力度再加大将对页岩气经济带来负面影响。

# 第7章　非常规油气开发宝典

## 7.1　油气井开发与监管

### 🔲 导语

　　油气井开发会给当地生态、文化遗址及现有土地的使用带来负面影响，甚至可能引发地震。虽然引发地震的可能性极小，但是大地只要发生轻微抖动就会破坏公众对钻井作业安全性的信任。非常规天然气开采需要社会支持，需要公众充分认识到环境和社会风险的可控性。因此，要有严格的监管制度来确保油气井设计、施工和生产符合规范。决策者、监管者和生产者以及其他人员可以遵从非常规油气井开发宝典，控制相关环境和社会影响。

### 🔲 课文

　　全球非常规天然气的开发前景主要取决于环境问题如何处理。为了使非常规资源开发带来更大的经济、环境和社会利益，全社会需要充分认识到环境和社会风险的可控性，支持非常规天然气开采。

　　决策者、监管者和生产者以及其他人员都可以根据下面的宝典处理环境和社会相关问题，从而获得社会对非常规资源开发的长期认可。

**谨慎选择井场**

　　油气钻井开发会影响当地社区及现有土地的使用，破坏文化遗址和生态，干扰居民生活，所以油气井选址要将这些影响减到最小。选址时要和当地有关人员和监管部门谨慎接洽，除需考虑地下地质条件外，还要考虑居住区、自然环境、当地生态、野生动物保护、道路和基础设施状况、水源和水处理问题、以及气候导致的季节性限制等。此阶段如果对地上多方面因素考虑周到的话，会减少或避免后期开发过程中的问题。

　　做好该地区的地质勘探，以便正确决定钻井和水力压裂方位，即对深部断层或其他地质特征是否会引发地震或使流体流过地层做出风险评估。计划详尽周到可以大大提高油气井产能和采收率，减少钻井数目，使水力压裂程度和环境影响降到最低。虽然引发地震的风险很小，但是轻微的大地抖动就足以破坏公众对钻井作业安全的信任。钻井前必须对该地区进行详细的地质研究，使生产者避开在有深部断层或其他高危地质特征的区域进行作业。

生产者也要仔细探查是否有旧井眼存在,或生油层上方浅层区域是否有天然甲烷存在,适当调整井位(或者井眼轨迹)以避开这些区域。

监测水力压裂过程,确保裂缝不超出产气层。页岩气和致密气生产使用的压裂液从产层渗漏到含水层的可能性很小,因为含水层位置很浅,但在某些例外情况下,理论上讲这种运移也是可能发生的。因此,要充分了解当地的地质情况,运用微地震(或其他)测量技术进行裂缝监测,使残余液造成的风险降至最低。

含气地层必须与井身穿过的其他地层(尤其是淡水层)完全分开,这是钻井工程基本准则,对井身结构设计、施工、固井以及综合测试都要做出严格的规定。要有规章制度来确保油气井设计、施工和操作达到完全隔离的效果。要采取多种措施防止泄漏,统一要求作业者系统地遵从推荐的行业标准做法。这种做法应该一直持续到井废弃,亦即贯穿于油气开发的全过程,包括生产结束以后的弃井过程。

要考虑实行水力压裂最小深度限制,让公众相信这个工程距地下水体足够远,不会产生威胁。除了采取一定措施确保油气井设计、施工和固井达到高标准外,监管者还要根据当地的地质状况和淡水层受污染的风险程度,限制页岩气井和致密气井的合理深度,如果超过了这个界限,就禁止应用水力压裂。

采取措施防止油气井地面溢流和泄漏,保证废液和废物妥善处理。这就需要所有从事钻井和生产作业的公司制定严格的规章制度,郑重承诺按照最高标准进行生产作业。要想防止事故发生,或者即使事故发生了也要将其影响控制在有限范围内,就必须具备良好的工作流程,进行人员培训,并且能及时启用溢流控制设备等。优化泥浆处理系统,以储罐和分离罐取代露天泥浆池(闭环系统),可以降低钻井过程中废弃物发生意外泄漏的风险。

常规油气开发一般遵循相当明确的程序,但是非常规开发的各个阶段之间没有明显的区分,通常是渐进式的。任何时候,生产商都会对其开发区块的某一部分进行勘探和评估,对另一部分进行开发,而还有的区域已经投入生产,每个阶段都要有不同的监管许可。非常规资源项目开发各阶段之间的界线模糊,使整个开发过程中生产商和监管方之间(生产商和当地社区之间)的交往更复杂。

例如,大多数司法管辖区的监管系统都要求勘探结束后提交一份详细的油气开发计划,通过后才能生产。然而,非常规开发摸索周期较长,在此阶段拿出完整的方案要难得多,而且过后哪怕只做一个小小的调整都可能会导致整个开发方案重新提交,重新批复,这对双方来说都很耗时且麻烦。

监管方主要关注油气井施工和设备安装是否规范,此外他们还得关注许多相关工程和油气井的长期影响,尤其是水的使用和处理,以及对未来水需求量等的评估,在土地使用、空气质量、交通和噪音等其他方面,也不可放松。一般情况下,监管系统如果侧重于逐井审批,而不是授权项目的话,就可能不会警示某些环境风险,从而错过减小风险的机会。例如,一些基础设施投资建设可能和单井无关,但是会总体减少大规模开采所造成的环境影响,如用于供水和排水的管网及污水处理中心。监管部门可以通过大范围发放许可证,并使其长期有效来促进这种投资。

在天然气开发过程中,政府通常帮助推动地区基础设施及时扩建,确保监管各部门鼓励或要求建设天然气运输能力,扩大当地电力供应。政府各级各部门之间必须密切协调和沟通,因为新兴行业的快速发展不仅给当地基础设施建设带来了压力,同时也给当地的社会服务带来了挑战。

## Key to Exercises

I.1. Each well site needs to be chosen based on the subsurface geology, but also taking into consideration populated areas, the natural environment and local ecology, existing infrastructure and access roads, water availability and disposal options and seasonal restrictions caused by climate or wildlife concerns.

2. We should assess the risk that deep faults or other geological features could generate earthquakes or permit fluids to pass between geological strata.

3. The regulator may choose to define an appropriate depth limitation for shale and tight gas wells, based on local geology and any risk of communication with freshwater aquifers, above which hydraulic fracturing is prohibited.

4. The blurred lines between the stages of an unconventional resource project.

5. Strong co-ordination and communication is necessary between different branches and levels of government, as the rapid growth of a new industry puts pressure not only on the local physical infrastructure, but also on local social services.

II.1. D  2. B  3. D  4. C  5. A  6. B  7. D  8. C  9. B  10. D

III.1. A  2. D  3. C  4. A  5. B

IV.1. 选址时要和当地有关人员和监管部门谨慎接洽。

2. 要充分了解当地的地质情况,运用微地震(或其他)测量技术进行裂缝监测,使残余液造成的风险降至最低。

3. 要考虑实行水力压裂最小深度限制,让公众相信这个工程距地下水体足够远。

4. 任何时候,生产商都会对其开发区块的某一部分进行勘探和评估,对另一部分进行开发,而还有的区域已经投入生产,每个阶段都要有不同的监管许可。

5. 常规油气开发一般遵循相当明确的程序,但是非常规开发的各个阶段之间没有明显的区分,通常是渐进式的。

Ⅴ.1. 计划详尽周到可以大大提高油气井产能和采收率,减少钻井数目、使水力压裂程度和环境影响降到最低。虽然引发地震的风险很小,但是轻微的大地抖动就足以破坏公众对钻井作业安全的信任。

2. 在天然气开发过程中,政府通常帮助推动地区基础设施及时扩建,确保监管各部门鼓励或要求建设天然气运输能力,扩大当地电力供应。政府各级各部门之间必须密切协调和沟通,因为新兴行业的快速发展不仅给当地基础设施建设带来了压力,同时也给当地的社会服务带来了挑战。

# 7.2 减少环境影响

## 导语

非常规油气开发不可避免地要占用很多土地,并对环境产生了重大影响,如污染物空气排放、井喷或溢流、压裂液泄漏以及污水排放造成的地下水污染等。压裂液含有危险物质,其返排液又增加了来自沉积物的重金属和放射物质等。而且地下水甲烷污染严重时会导致居民楼爆炸,氯化钾可致使饮用水盐渍化,这些现象在油气井附近区域有报道。未来生产与开发的关键在于可持续发展,环境影响对政府以及监管者构成了严峻的挑战。

## 课文

**妥善处理水**

提高生产效率,减少淡水消耗;尽可能循环使用水资源,减轻当地用水压力。为了鼓励生产人员循环用水,提高水资源利用效率,需要制定页岩气、致密气和煤层气开发规章制度。用水量最大的是水力压裂过程:考虑到规模经济效益,应该尽可能地循环利用压裂作业中产生的大量返排水,减少卡车运水、水供应和废水处理等问题和成本。

安全储存及处理采出水和废水。要针对废水的安全储存,制定严格、系统的规章制度。采取措施确保施工牢固,确保露天钻井液池不渗漏,最好使用储罐。现有技术可以将采出水和废水以不同成本处理成各种标准。监管者要根据当地具体情况,制定适当的标准(包括淡水供应及废水处理方法选择),并且监督执行。生产者要始终按照不断提升的标准对生产负责。使用处理的循环水可能会带来多方面的长远利益,而将废水排放至各路水道或者蒸发池将产生更大的社会和环境成本,因此生产商以最低成本生产并不一定是最好、最经济的办法。

尽量少使用化学添加剂,开发使用更环保的替代品。压裂液添加剂的披露应不与不断创新的激励机制相矛盾。相关行业应当致力于压裂液开发,如果压裂液不慎泄漏也不破坏地下水质,或者采用其他技术减少化学添加剂的使用。

### 杜绝天然气排放、减少天然气燃烧排放及其他物质的排放

在完井过程中,以天然气零排放和燃烧排放最小化为目标,减少整个生产过程中温室气体的不稳定排放。最好的做法是将完井期间所产生的天然气回收并销售,而且公共管理部门要制定各种相关条例限制天然气排放与燃烧,并提出安装最小化气体排放设备的具体要求。这些措施也会同时降低挥发性有机化合物等常规污染物的排放。生产商应当把排放指标的制定视为其总体战略方针的一部分,使公众相信他们在努力减小其生产活动所带来的环境影响,并且将原本打算排放或燃烧掉的天然气卖出,获取一定的经济效益。天然气整个产业,包括常规天然气生产商和中下游生产公司,都应表示他们同样关心生产阶段以外的甲烷排放,例如运输与配送环节等。

最大程度地减少车辆、钻机、泵与压缩机等对空气的污染。车辆及设备污染通常受制于环境与能效标准(政府负责制定相关标准)。生产商和服务商都应当认识到使用现有最清洁的车辆和设备的优势,例如,电动车与气动钻机可以降低当地的空气与噪声污染。

### 从大处着眼

力争实现规模经济以及当地基础设施的协调发展,减少对环境的影响。为减少环境影响而进行的基础设施投资,对单井生产来说也许不能获取利润,但对大规模发展是有利的。良好的监管可以确保对特许开发区及相关基础设施进行合理的空间规划(如进入施工现场的通道、水资源及污水处理设备、天然气加工设备、增压站和管线等),从而有利于实现收益。建立公共设施走廊和多功能道路的想法有助于集中建设基础设施,避免对环境造成更大的影响。生产商可以从多方面获益,比如用一个钻井平台钻多底井(水平钻井通向油藏的不同方位):这种做法可能对井场附近破坏比较大,但可

大量减少对地面环境更大范围的破坏。另一个例子是建设水运管网，这虽然需要限期投入，但却可以省去施工期间上千辆卡车的运输，也可以缩减设备成本。生产商做出良好的工程和后勤供应计划还不够，还需要有前期的战略评估以及公共管理部门的及时管理。

考虑丛式钻井、开采和运输等生产活动对环境造成日积月累的区域性影响，尤其是水资源利用和污水处理、土地使用、空气质量、交通和噪声等方面。所有油气资源开发都需要进行大量基础设施建设、材料供应、钻井、开采、加工和销售等活动。而非常规油气开发尤甚，因为有更多的井要钻。因此，单井开发过程中的道路运输量，土地和水的用量，或钻井过程产生的噪声等也许还可以忍受，但随着生产规模的扩大，生产活动量就要增加。监管者需要评估这些影响的累积效应，并且给与适当反馈。在用水需求上，区域评估尤其重要。

水资源与水污染风险对政府以及监管者构成了严峻的挑战。只有进行严格的数据采集、对用水需求量（页岩和致密气）进行评估与监控，对采出水（煤层气）和废水（所有油气）的水质进行检测，才能做出合理的决定。

## Key to Exercises

I. 1. Rigorous and consistent regulations are needed to cover safe storage of waste water, with measures to ensure the robust construction and lining of open pits or, preferably, the use of storage tanks.

2. Fracturing fluid additives should be disclosed. The industry should commit to the development of fluid mixtures that, if they inadvertently migrate or spill, do not impair groundwater quality, or adopt techniques that reduce the need to use chemical additives.

3. Best practice is to recover and market gas produced during the completion phase of a well, and public authorities need to consider imposing restrictions on venting and flaring and specific requirements for installing equipment to help minimise emissions.

4. Pollution from vehicles and equipment is often controlled by existing environmental andfuel-efficiency standards.

5. The construction of a pipeline network for water requires upfront investment but obviates the need for many thousands of truck movements over the duration of a project and can lower unit costs.

II. 1. A   2. D   3. B   4. B   5. C   6. C   7. A   8. A   9. B   10. C

Ⅲ.1. A  2. D  3. B  4. B  5. A

Ⅳ.1. 使用处理的循环水可能会带来多方面的长远利益,而将废水排放全各路水道或者蒸发池将产生更广的社会和环境成本,因此生产商以最低成本生产并不一定是最好、最经济的办法。

2. 相关行业应当致力于压裂液开发,如果压裂液不慎泄漏,不要破坏地下水质,或者采用其他技术减少化学添加剂的使用。

3. 天然气整个产业,包括常规天然气生产商和中下游生产公司,都应表示他们同样关心生产阶段以外的甲烷排放,如运输与配送环节等。

4. 车辆及设备污染通常受制于环境与能效标准(政府负责制定相关标准)。

5. 所有油气资源开发都需要进行大量基础设施建设、材料供应、钻井、开采、加工和销售等活动。

Ⅴ.1. 为了鼓励生产人员循环用水,提高水资源利用效率,需要制定页岩气、致密气和煤层气开发规章制度。用水量最大的是水力压裂过程:考虑到规模经济效益,应该切实可行地循环利用压裂作业中产生的大量返排水,减少卡车运水、水供应和废水处理等问题和成本。

2. 水资源与水污染风险对政府以及监管者构成了严峻的挑战。只有进行严格的数据采集、对用水需求量(页岩和致密气)进行评估与监控,对采出水(煤层气)和废水(所有油气)的水质进行检测,才能做出合理的决定。

# 7.3  社会责任

▢ 导语

　　每个人或每个组织都要为维持经济与生态之间的平衡尽自己的义务,这就是社会责任。经济发展(物质意义上的)总是要以社会和环境的福祉为代价。社会责任即意味着维持两者之间的平衡。公司社会责任的作用是建立内部的自我管理机制,使企业确保遵守法律精神、道德标准以及国际准则。公共管理部门也要严格限制生产活动对环境的破坏。油气产业的可持续性发展需要政府、业界以及个体经营者共同行动起来,赢得并维持公众的认可。

▢ 课文

**确保环保始终不打折扣**

　　在勘探之前就开始坚持在每个开发阶段都要和当地社区、居民和其他股东接洽合作,为其提供充分机会,对开发计划和生产活动广开言路,倾听公众关心的问题并及时做出适当反馈。

仅仅向公众提供信息是不够的;业界和公共管理部门还需要与当地社区和其他股东接洽,告知实际情况并求得认可,而公众往往会对公司的后续生产提出批评性建议。生产商要开诚布公地说明其生产情况和环境、安全和健康风险,以及如何解决这些问题。公众应该对发展中有何挑战、风险和利益等一清二楚。公共管理部门的主要职能则是提供可靠的、科学性的背景信息,利用这些信息可以进行合理辩论,并激励股东之间的携手合作。

在生产活动开始之前,就要对关键性环境指标设立基准线,例如地下水质量,在生产过程中也要进行持续监测。这是监管部门、业界和其他股东们共同的责任。收集的数据要公开,所有的股东都应有机会解决大家提出的疑虑问题,这是赢得公众信任所必需做到的。在新井钻探之前,至少资源管理及监管机构必须掌握地下水质信息(如果是煤层气开采,则需要掌握地下水位信息),以便于设定基准线,依照这个基准线就可以比较水位和水质发生的变化。

在对压裂液添加剂及其用量进行强制性完全披露的同时,还要对用水量、废水量及其特性,以及甲烷和其他空气排放物等方面的生产经营数据进行测量并披露。数据真实、透明、测量准确对树立公众信心是非常重要的。例如,为了推动并确保废水经过适当处理后排放,就必须对污水进行有效跟踪和记录。然而不愿意透露压裂过程中所使用的化学品及其用量,虽然从商业竞争方面来说可以理解,但却会迅速滋生当地居民和环保团体的不信任。

将生产过程中的破坏降至最低,放眼于社会和环境责任,并确保当地社区也能分享到经济利益。现有的法律和法规通常要求生产经营者的活动要对环境和社会负责,然而,生产经营者不能仅限于满足法律要求的最低线,而要表明其致力于当地经济的发展和环境保护,比如注意当地居民对卡车运输量和时间安排方面的意见。特别是在矿权归州所有的管辖区内(而不像美国的某些地区,地面土地所有者也可能是地下矿权持有者,有权获得特许使用费),必须要让生产区居民获得实实在在的好处。然而这很难及时兑现,因为从工程项目开始到赚取第一桶金要有一段延迟时间,无论对政府、矿权拥有者,还是生产方都是如此。但是如果管理部门和开发商能极早公开承诺其将随着勘探和生产进程,扩建基础设施和服务设施,便能够起到一定作用。政府部门应愿意拿出部分收入(税收、特许使用费等)投资于有问题地区的发展。

确保以相应的资源、对稳健的监管制度政治上的适当支持、足够的审计认证人员以及可靠的公共信息使非常规天然气产量达到预期水平。政府的一个重要关注点应该放在确保在非常规天然气开发的环境和技术方面有足够的知识储备,确保高质量数据的获得,确保正确的科学运用和推广。要有足够多的主动积极、技术过硬和资金充足的监管者,他们是进行负责任的非常规资源开发必不可少的力量。

在规定性监管和基于绩效的监管之间找到政策制定的适当平衡点,以便在保证生产高标准的同时也能促进创新和技术进步。在某些生产中,必须有详细的规定和检查才能保证环保达标;但是在技术日新月异的发展情况下,要对生产过程的方方面面都进行控制有时是不可能的,也是不可取的。设定工作标准,让生产人员自己找到最合适的方式去达标,常常能比规定性方法产生更好的效果。例如对水利用改进方面设定了最低标准,或者是要求检测最佳水泥质量。这对施工方是一种压力,他们必须证明使用了最佳水泥。

无论选择哪种方法或是哪几种方法,如果没有通过认证,都要严格、强制执行和接受处罚,最终可能吊销施工执照。

确保针对各级风险的应急方案健全有效。生产商和当地应急服务部门必须有健全的成文应急方案和程序,包括人员培训和应急设备等,以便迅速有效地应对意外事故。

不断改进规章制度和生产方法。技术在不断进步,方法在不断更新。监管清晰和稳定固然是优势,但政府也必须积极将不断变化的产业部门中的经验教训整合进监管法规。对于行业来说,追求最佳做法就意味着要不断积极提高标准,并提供条件来达到这些标准。

认可对环保工作的独立评估和验证。如果第三方对企业业绩的认证真实可信的话,就会有力地赢得和维持公众的认同,同时也有助于公司践行最佳方法。这些独立评估应该来自于受公众信任的机构,无论是学术或研究机构或是独立的监管及认证机构。

这些黄金规则的应用,需要政府和行业共同行动起来。虽然维持公众信心的最终责任落在业界,但是政府应该设置监管构架,公布必要规则,并通过组织科研等活动提供支持。试图在全球统一确定政府、天然气生产商和其他个体经营者的具体角色,是不现实的事情。各国的法律、地质、社会政治背景、土地利用、水资源和其他许多方面的条件都有所不同。

## Key to Exercises

Ⅰ.1. Operators need to explain openly and honestly their production practices, the environmental, safety, and health risks and how they are addressed. The public needs to gain a clear understanding of the challenges, risks and benefits associated with the development.

2. Good data, measurement and transparency are vital to public confidence.

3. The primary role of the public authorities is to provide credible, science-based background information that can underpin an informed debate and provide the necessary stimulus for joint endeavour between the stakeholders.

4. Existing legislation and regulations usually require operators to act in an environmentally and socially responsible manner.

5. Because technology is moving rapidly.

Ⅱ. 1. A    2. B    3. B    4. D    5. C    6. B    7. B    8. D    9. B    10. B

Ⅲ. 1. C    2. B    3. A    4. D    5. C

Ⅳ. 1. 仅仅向公众提供信息是不够的;业界和公共管理部门还需要与当地社区和其他股东接洽,告知实际情况,并求得认可,而公众往往会对公司的后续生产提出批评性建议。

2. 公共管理部门的主要职能则是提供可靠的、科学性的背景信息,利用这些信息可以进行合理辩论,并激励股东之间的携手合作。

3. 收集的数据要公开,所有的股东都应有机会解决大家提出的疑虑问题,这是赢得公众信任所必需做到的。

4. 然而这很难及时兑现,因为从工程项目开始到赚取第一桶金要有一段延迟时间,无论对政府、矿权拥有者,还是生产方都是如此。

5. 政府的一个重要关注点应该放在确保在非常规天然气开发的环境和技术方面有足够的知识储备,确保高质量数据的获得,确保正确的科学运用和推广。

Ⅴ. 1. 数据真实、透明,测量准确对树立公众信心是非常重要的。例如,为了推动并确保废水经过适当处理后排放,就必须对污水进行有效跟踪和记录。然而不愿意透露压裂过程中所使用的化学品及其用量,虽然从商业竞争方面来说可以理解,但却会迅速滋生当地居民和环保团体的不信任。

2. 这些黄金规则的应用,需要政府和行业都共同行动起来。虽然维持公众信心的最终责任落在业界,但是政府应该设置监管构架,公布必要规则,并通过组织科研等活动提供支持。试图在全球统一确定政府、天然气生产商和其他个体经营者的具体角色,是不现实的事情。各国的法律、地质、社会政治背景、土地利用、水资源和其他许多方面的条件都有所不同。

# References

［1］ Akash Shah, Shannon Shuflat. Global Shale Gas Technologies and Markets. ⌊2011 −02⌋. http://www. sbireports. com.

［2］ Michael Ratner, Mary Tiemann. An Overview of Unconventional Oil and Natural Gas: Resources and Federal Actions. ［2014 −01］. http://www. fas. org/sgp/crs/misc/R43148. pdf.

［3］ Deborah Gordon. Understanding Conventional Oil. ［2012 − 05］. http://carnegieendowment. org/files/unconventional_oil. pdf.

［4］ International Energy Agency, World Energy Outlook. ［2011］. http://www. iea. org/weo/docs/weo2011/executive_summary. pdf.

［5］ Gabriel Chan, John M. Reilly, Sergey Paltsev, and Y. H. Henry Chen, Canada's Bitumen Industry Under $CO_2$ Constraints MIT, Report 183, http://globalchange. mit. edu/files/document/MITJPSPGC_Rpt183. pdf.

［6］ Rebecca C. Rooney, Suzanne E. Bayley, and David W. Schindler, Oil Sands Mining And Reclamation Cause Massive Loss Of Peatland And Stored Carbon. PNAS, November 3, 2011.

［7］ Sarah M. Jordaan, David W. Keith and Brad Stelfox. Quantifying Land Use Of Oil Sands Production: A Life Cycle Perspective. Environmental Research Letters, volume 4. ［2009 − 05 − 12］. http://iopscience. iop. org/1748 − 9326/4/2/024004/fulltext.

［8］ Energy Technology Systems Analysis Program. Unconventional Oil & Gas Production. IEA ETSAP-Technology Brief, 2010:02［2010 − 05］. http://www. etsap. org.

［9］ Reeves, Scott R. , George J. Koperna, and Vello A. Kuuskraa. Technology, efficiencies keys to resource expansion. Oil and Gas Journal. October 1, 2007.

［10］ Warpinski, N. R. , J. E. Uhl, B. P. Engler, J. C. Lorenz, and C. J. Young. Development of Stimulation Diagnostic Technology: Annual Report. GRI − 97/0115. Sandia National Laboratories, 1997 ［2013 − 01 − 29］. http://prod. sandia. gov/techlib/access-control. cgi/1997/971327. pdf.

［11］ Harbert, Karen A. Tapping America's Unconventional Oil Resources for Job Creation and affordable Domestic Energy: Technology & Policy Pathways. Testimony to the US House of Representatives Committee on Science, Space, & Technology. April 17, 2012.

［12］ Neal, Howard. Oil and Gas Technology Development. National Petroleum Council Topic, 2007:26［2013 − 01 − 14］. http://downloadcenter. connectlive. com/events/npc071807/pdf-downloads/Study_Topic_Papers/26 − TTG − OGTechDevelopment. pdf.